Structural Failure in Residential Buildings

Volume 1   Flat Roofs, Roof Terraces, Balconies

The publishers are grateful for the help and advice of John G. Roberts B.Arch., ARIBA, of the Welsh School of Architecture, University of Wales Institute of Science and Technology, in the preparation of the English-language edition of this book.

Structural Failure in Residential Buildings

Volume 1
Flat Roofs, Roof Terraces, Balconies

Erich Schild
Rainer Oswald
Dietmar Rogier
Hans Schweikert

Illustrations by Volker Schnapauff

A HALSTED PRESS BOOK

JOHN WILEY & SONS
New York

Published in the U.S.A.
by Halsted Press, a Division of
John Wiley & Sons, Inc., New York
First published in 1976 in the Federal Republic of Germany
by Bauverlag GmbH, Wiesbaden and Berlin
English edition first published in Great Britain 1978
by Crosby Lockwood Staples
Frogmore, St Albans, Hertfordshire AL2 2NF and
3 Upper James Street, London W1R4BP

Library of Congress Cataloging in Publication Data
Structural failure in residential buildings.
    Translation of Bauschadensverhütung im Wohnungsbau.
    'A Halsted Press book.'
    Includes bibliographical references.
    CONTENTS: v. 1. Flat roofs, roof terraces,
balconies.
    1. Building failures.   2. Dwellings.   I. Schild,
Erich, 1917–
TH441.B34   1978   690'.1'5   77-28647
ISBN 0-470-26305-9 (v. 1)

Copyright © 1978 by Crosby Lockwood Staples

Translated from the German by Sheila Bacon

Printed in Great Britain by
Fletcher & Son Ltd, Norwich

This publication is the result of the research project 'Problems of structural failure — prevention of structural failure in residential buildings', commissioned by the Ministry of the Interior (Residential Buildings Department) of North Rhein-Westphalia and carried out at the Technical University of Aachen, Faculty of Building Construction III — Building Science and Problems of Structural Failure (senior professor Dr-Ing. Erich Schild).
Research team:

Erich Schild	Prof. Dr-Ing.
Rainer Oswald	Dipl.-Ing.
Dietmar Rogier	Dipl.-Ing.
Volker Schnapauff	Dipl.-Ing.
Hans Schweikert	Dipl.-Ing.
Christian Lange	cand. arch.
Jürgen Roder	cand. arch.
Klaus-Dieter Weiss	cand. arch.
Axel Werner	cand. arch.

Norma Gottstein
Gunda Hoppe

Preface to English edition

We are now reaping with alarming alacrity the harvest of building failures of the post-war building boom where inappropriate use was made of materials without a proper understanding of their properties or their performance in varying conditions. This publication is based on research carried out in Western Germany but its principal findings are applicable to work undertaken in the United Kingdom although our climatic polarities are not so severe.

Many failures are now occurring in modern residential buildings which illustrate defects in detailing and constructional techniques, lack of site supervision and subsequent maintenance. Their repair is extremely costly mainly because of their rigid frame construction.

Conformity with building regulations and bye-laws does not guarantee a defect-free building in all respects; indeed new buildings are not risk-free and it is unlikely that failures can be avoided completely.

Building methods developed during the past twenty years under intense pressure of time and cost without adequate testing resources inevitably lead to disasters. If this problem is to be overcome there should be strict observance of the recommended Code of Practice and appropriate use of materials.

Physical causes of defects and details known to be suspect must not be repeated and technical reliability must be improved. Practitioners must obtain 'feed-back' from completed projects to ensure the viability of their design. Defects must be correctly diagnosed and the causes and related factors investigated. There may be more than one cause but in most instances it is possible to identify the primary one.

Architects must investigate and understand the discipline that multi-layer construction imposes on design solutions. This publication offers recommended solutions to the problems inherent in flat roofs, roof terraces and balconies. These are the areas which possibly have the highest number of failures in residential buildings.

John G. Roberts
WSA UWIST
Cardiff, 1977

Contents

C Roof terraces

C 1 *Typical cross-section*

C 2 *Points of detail*

D Balconies

D 1 *Typical cross-section*

D 2 *Points of detail*

Preface

This publication represents the final report of an investigation into the most important causes of structural damage to roofs, roof terraces and balconies. The investigation was carried out within the framework of the research project 'Problems of structural failure — prevention of structural damage in residential buildings'. Gratitude is due to the client, the Ministry of the Interior (Residential Building Department) of North Rhein-Westphalia and Prof. Dr-Ing. H. Paschen, Prof. Dr-Ing. W. Triebel and Dr-Ing. H. Künzel for their support of the project.

This report was preceded by two working reports. In the first phase the extent and crucial areas of structural damage to residential buildings were investigated by means of representative enquiries into the condition of new buildings in North Rhein-Westphalia. This resulted in an intensive investigation of structural damage to roofs, roof terraces and balconies:

1. An enquiry was undertaken among publicly appointed and qualified structural specialists in North Rhein-Westphalia on roofs, roof terraces and balconies.

   In order to achieve a random selection and thereby a statistically representative result, the experts reported on the two most recent cases of damage they had worked on, on a standard questionnaire.

   This enquiry attained special significance with respect to the viability of statistical evidence. It gave a scientific foundation for establishing the quantitative distribution of damage to the constructions investigated and determined the special problem areas and weak points in the design of roofs, roof terraces and balconies which were used in residential buildings.

2. An evaluation of the literature was undertaken, including product information from industrial manufacturers.

3. The findings were collated of the specialist investigation by the researchers of more than two hundred cases of damage to roofs, roof terraces and balconies.

Thanks are due in this connection to the specialists for their co-operation. Approximately 62% of those questioned replied to the enquiry; of those experts in North Rhein-Westphalia concerned with damage to roofs, about 45% participated in the detailed enquiry by completing the questionnaire.

The final evaluation of the examples of defects described in detail by the specialists gave an objective picture of the susceptibility to damage of various structural components and construction. Thus the result of the first enquiry was confirmed: that flat roofs, and especially those with no falls, are more susceptible to failure than roofs with a fall. Approximately half the damage to flat roofs occurred to those which had no fall.

Frequently, defects in detailing led to damage. A surprising result of the enquiry was that failure to construction details of parts of the building — to connections, joints, openings, etc. — occurred far more frequently than damage to the typical cross-section: 72·5% of all reports of one or more instance of failure referred to defective areas of detail. Damage to the typical cross-section was observed only in 57% of the experts' reports.

The vertical edge of the structural member was a particular problem area in the structures investigated: 37% of all failure occurred here. A high proportion of damage was also recorded to abutment junctions where the flashings were not taken sufficiently far up the wall.

A further important result of the investigation was that, in contrast to the general belief, both double slab ventilated flat roofs and solid slab warm roofs are subject to defects (47% damage to solid slab, 53% to double slab roof structures). Damage caused by condensation as a result of defective or insufficient ventilation forms a particularly high proportion.

Finally, damage in the bearing of the roof edges represented a further area of failure. Damage to these areas occurred in over 40% of the solid slab and double slab roofs of reinforced concrete that were investigated.

This book, *Structural Failure in Residential Buildings, Vol. 1 — Flat roofs, Roof terraces, Balconies*, is the result of the third phase of the investigation and the aim of the research project. It is directed at those concerned with the designing and construction of these structural elements — clients, architects, engineers, site staff, industrial construction manufacturers, training establishments and all others connected with the building industry.

Following the representative research into structural failure, this report develops the two methods of work described, and deals with

(a) problem areas in materials and
(b) construction methods and structural procedures used extensively in residential building at the present time.

Methods of construction which, according to statistical evidence, are rarely defective either because they have no marked defects or because they are used in residential building only to a limited extent, are not dealt with here. Accordingly this book does not represent a complete building construction manual on roofs, roof terraces and balconies. This was not the intention. The urgent task of this work is to prevent those defects that have been found to occur frequently and to interpret statistically valid research results to discover the crucial areas.

By arrangement and composition this report is intended for direct use as a working basis in the design, testing and manufacture of structural components. In order to allow information to be found easily, it is arranged in a uniform system of classification.

Each of the four types of structural component

A Solid slab flat roofs
B Double slab flat roofs
C Roof terraces
D Balconies, loggias, arcades

is dealt with independently and separately. The different aspects of individual types of structural component (e.g. connection to abutments) are presented as separate problems. Treatment of the individual points of defect associated with a problem is purposely not restricted to the most simple solution. The recommendations for the avoidance of a defect are preceded by a description of the damage for which the suggested remedy is prescribed; this is followed by an outline of the implications of the damage analysis

SCHEMATIC REPRESENTATION OF STAGES OF WORK

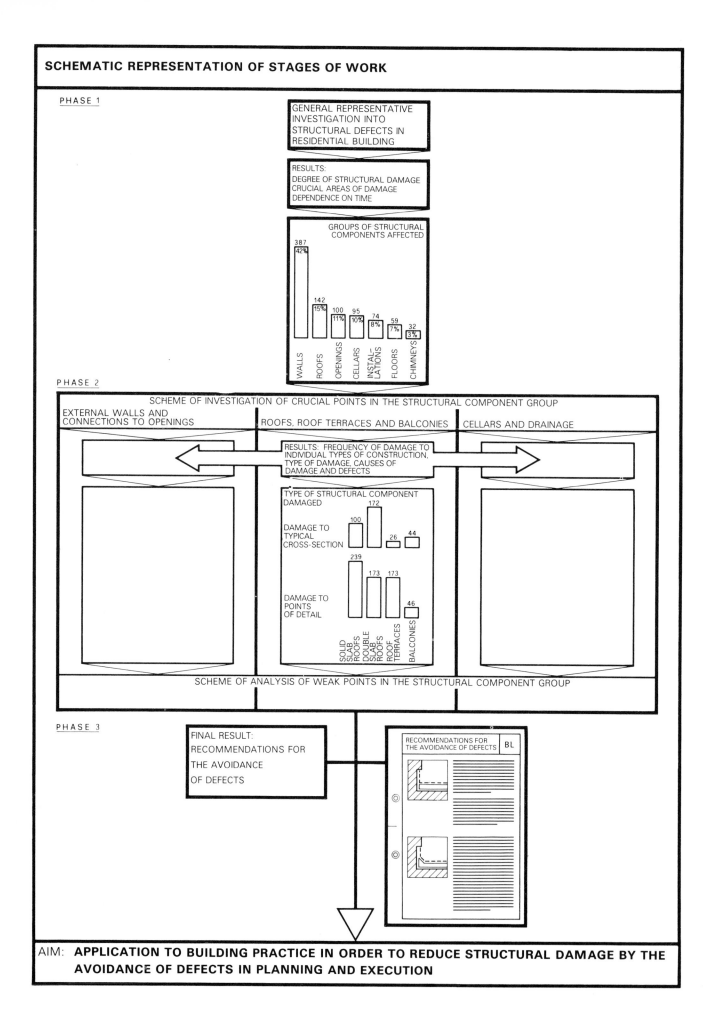

PHASE 1

GENERAL REPRESENTATIVE INVESTIGATION INTO STRUCTURAL DEFECTS IN RESIDENTIAL BUILDING

RESULTS:
DEGREE OF STRUCTURAL DAMAGE
CRUCIAL AREAS OF DAMAGE
DEPENDENCE ON TIME

GROUPS OF STRUCTURAL COMPONENTS AFFECTED

387 42% WALLS
142 15% ROOFS
100 11% OPENINGS
95 10% CELLARS
74 8% INSTAL-LATIONS
59 7% FLOORS
32 3% CHIMNEYS

PHASE 2

SCHEME OF INVESTIGATION OF CRUCIAL POINTS IN THE STRUCTURAL COMPONENT GROUP

EXTERNAL WALLS AND CONNECTIONS TO OPENINGS

ROOFS, ROOF TERRACES AND BALCONIES

CELLARS AND DRAINAGE

RESULTS: FREQUENCY OF DAMAGE TO INDIVIDUAL TYPES OF CONSTRUCTION, TYPE OF DAMAGE, CAUSES OF DAMAGE AND DEFECTS

TYPE OF STRUCTURAL COMPONENT DAMAGED

DAMAGE TO TYPICAL CROSS-SECTION
100
172
26
44

DAMAGE TO POINTS OF DETAIL
239
173
173
46
SOLID SLAB ROOFS
DOUBLE SLAB ROOFS
ROOF TERRACES
BALCONIES

SCHEME OF ANALYSIS OF WEAK POINTS IN THE STRUCTURAL COMPONENT GROUP

PHASE 3

FINAL RESULT:
RECOMMENDATIONS FOR
THE AVOIDANCE
OF DEFECTS

RECOMMENDATIONS FOR THE AVOIDANCE OF DEFECTS | BL

AIM: **APPLICATION TO BUILDING PRACTICE IN ORDER TO REDUCE STRUCTURAL DAMAGE BY THE AVOIDANCE OF DEFECTS IN PLANNING AND EXECUTION**

with regard to physically or technically related parts of the structure.

A change in current attitudes can be achieved only if the logical basis of the recommendations is clearly understood, and this, since it comprises a fundamental awareness of the problem, allows the recommendations to be applied to constantly changing situations. The basic data merit the close attention of the reader. For easily accessible information, the most important points are summarised in the notes to each section. The bibliography allows the reader to pursue further problems of specific interest.

Details of defects, examples of damage and correct structural solutions are shown in diagrams. To emphasise the general problem, location sections are to a smaller scale than the detailed sections. Complex structural solutions are shown but the diagrams should not be regarded as construction drawings, and represent only one of several possible solutions; they do, however, attempt to adhere to basic construction principles. Direct recommendation or rejection of proprietary products has been avoided, as the intention is to present principles and rules of construction that are as universally applicable as possible. Nevertheless, the recommendations provide a means of assessing available building materials, components and construction methods before a choice is made.

Although the constructional research for the project was carried out in 1974, this final report takes into account more recent technical developments that have been published or have otherwise come to the knowledge of the authors in the course of their work as practising engineers.

To specify a solution for each type of failure demanded some courage: however, the ensuing and inevitable criticism is one of the aims of this book, in that it will promote a lively exchange of views between those active in the field.

Following this work on roofs, roof terraces and balconies, research is now being carried out, and is planned for similar publication, on external walls and openings, basements and drainage.

Problem: Sequence of layers and individual layers

The solid slab flat roof has to accommodate various stresses and design variabilities yet still provide a functional roof structure that is slender in depth and low in weight. Characteristic of this is the build-up of several layers of different materials in a multi-layer unit. The functions of the individual layers are basically determined by the demands and stresses to which the roof is subjected as an external component. To achieve this the following important functions must be fulfilled:

(i)   protection against rainwater penetration to the interior of the building and to the roof structure itself;

(ii) heat insulation (winter and summer);

(iii) loadbearing capacity (dead and imposed loads).

The permutations of these functional layers identify the constructional type of solid slab flat roofs. These can be identified as a warm roof, an 'inverted roof' and a waterproof concrete roof.

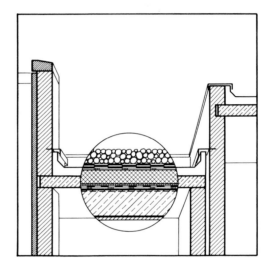

The individual functional layers may be composed of a variety of materials and products. Of the construction types listed for the solid slab flat roof in its various forms, the one most commonly used in housing is that of the 'warm roof' made of solid reinforced concrete covered with bituminous roofing felt.

In a statistical survey of structural defects, damage to this type of flat roof construction formed about 84% of the total. Suggestions for preventing failure in this type of construction therefore form the basis of the following recommendations.

The structural failures which occur in the typical cross-section demonstrate clear key areas of damage. On the one hand there are defects due to damp resulting from leaks and incorrect positioning or dimensioning of the thermal insulation and the vapour barrier in terms of the dispersion of water vapour; on the other hand, distortion and cracking in the wall plates and supporting structures occur because no provision has been made for movement and strong thermal stresses within the structural roof itself.

On the following pages the defects of the solid slab and the typical recurring errors in design and execution are illustrated and analysed. From these, conclusions are drawn for the avoidance of these failures.

Solid slab flat roof

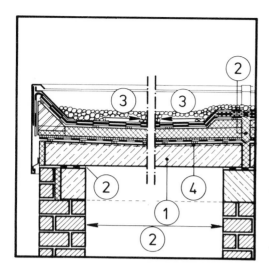

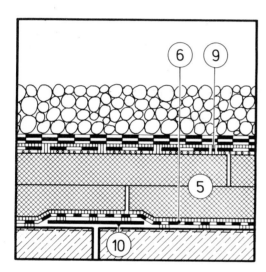

1 The loadbearing roof surface or the distance between supports must be dimensioned so that bending under increased load due to traffic, gravel or standing water does not lead to unacceptable stressing of the roof or to changes in the fall of the roof (see A 1.1.2).

2 Solid roof surfaces over 12 m in length must have expansion joints at least 20 mm wide unless the expected changes in length are proved by calculation to be harmless (see CP 110: Pt. 1: App. B) or the design of the bearing allows the roof slabs to slide freely (see A 1.1.3).

3 The waterproof layer must show $\geqslant 3\%$ (approx. $1 \cdot 4°$) slope to the gutters, especially if no upper surface protection layer is provided to control dampness (see A 1.1.8).

4 The slope should be achieved (a) by the structural roof being cast with an inbuilt fall, (b) by forming a lightweight screed laid to falls or (c) by laying thermal insulation on the structural roof to form the slope (e.g. wedge-shaped insulating slabs, bitumen-bonded granulated insulating material). (See A 1.1.4 and 1.1.8.)

5 Thermal insulating layers should have a minimum thermal resistance of $1 \cdot 3$ m² K/W and should always be placed above the loadbearing roof slabs. If this is not possible, the thermal insulation above the vapour barrier should be strengthened. The larger thermally conditioned movements of the roof slabs have to be allowed for in the construction of expansion joints and roof supports (see A 1.1.3 and 1.1.4).

6 Where a solid slab flat roof has an external vapour barrier, this should have a vapour seal value (diffusion resistance equivalent air layer thickness) of $\geqslant 100$ m and should be placed under the thermal insulation (see A 1.1.4).

7 In order to avoid defects, insulating layers should if possible be double-layer with concealed joints or single-layer with overlaps (see A 1.1.5).

8 Humidity between the roof membrane and vapour barrier should be avoided. The thermal insulation materials should therefore be protected from damp during storage and erection (see A 1.1.5 and 1.1.7).

9 Between the roof membrane and the thermal insulating layer there should be a stabilising layer forming a channel and allowing spot or strip adhesion (see A 1.1.7).

10 In the case of large structural slabs the adjoining layer should be left free of adhesive over the expansion joints of the slabs (transverse joints) to a width of 300 mm (see A 1.1.7).

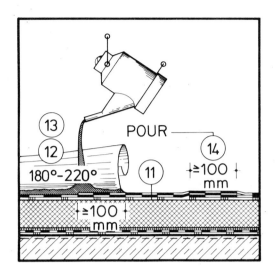

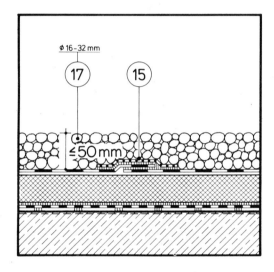

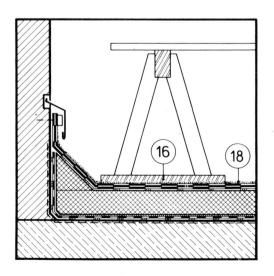

11 The nature of the base must permit a methodical laying of the sealing compounds. It must be flat, dry, not too porous and be free of oil and paint residue (see A 1.1.9).

12 Heat sealing – 180–220°C according to the type of bitumen used – should be carried out only in dry weather and at outside temperatures $\geq +5°C$ (see A 1.1.10).

13 In the selection of the sealant or insulating material, care should be taken to achieve good adhesion, while avoiding thermal or chemical conditions that could prove destructive to the insulating layer (see A 1.1.5 and 1.1.9).

14 In the case of sealing layers made of bituminous roofing sheets the longitudinal and transverse overlap seams should be ≥ 100 mm wide. The sheets should be saturated and bonded down over the whole surface. The pouring and rolling process is basically preferable to other adhesion techniques (see A 1.1.9).

15 Single-layer loose-laid synthetic roofing sheets should, as well as bonding or welding of the overlap seams, receive additional protection in the form of covering strips or an in situ bitumen dressing (see A 1.1.9).

16 Necessary work on roof components in the course of erection, during which the sealing layer is often unprotected, should only be carried out using the appropriate precautionary measures. Plank or board supports prevent pressure points from trestles and damage from movement of the gravel (see A 1.1.10).

17 In principle, every flat roof should have an upper surface protective layer in the form of loose gravel; this should be practicable with a roof slope of $\leq 5°$. It should be ≥ 50 mm deep, with a coarseness of 16–32 mm dia (round grain), and be freely and equally distributed. The weight of the gravel layer must be taken into account in the calculation and selection of material for the roof structure (see A 1.1.8 and 1.1.11).

18 Thin upper surface protective layers (reflective strips, sandings, etc.) must be inspected and, if necessary, renewed at frequent intervals to check that they are functioning correctly (see A 1.1.11).

Solid slab flat roof
Sequence of layers and individual layers

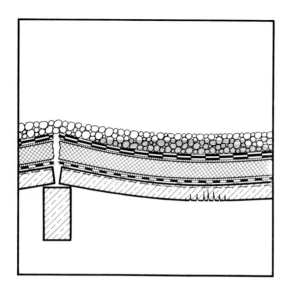

Marked sagging often occurs on the loadbearing parts of the roof, causing cracks to form in the solid roof slabs or damage to the roof structure. In light roofs made of steel lattice beams or timber joists, cracks occur in the sealing layer in the area of the intermediate bearings. Where a roof surface is designed with very little or no fall, puddles appear in the hollows formed. In the case of cantilevered roof slabs with minimal roof-edge slope, sagging leads to overflowing of water over the edge of the roof.

Points for consideration

— If roof structures, in relation to loading (dead and imposed loads) and the distance between supports, are dimensioned in such a way that sagging occurs, the efficiency of the roof will be impaired by the formation of cracks and puddles — especially where the roof has no fall.
— Similar situations are found in reinforced concrete roof slabs when the reinforcement does not relate to the actual load-transfer points (e.g. unintended bearing on partition walls).

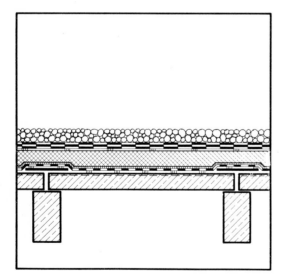

Recommendations for avoidance of defects

● The structural roof and distance between supports must be dimensioned so that sagging under additional loading due to traffic, gravel or standing water does not lead to unacceptable stresses in the roof or to changes in the fall of the roof.

● Unintended loading on partition walls is to be avoided by use of an appropriate separating joint.

Solid slab flat roof
Sequence of layers and individual layers

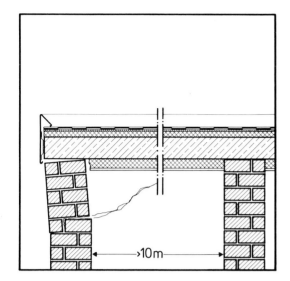

Where rigid roof coverings have structural members longer than 12 m and external thermal insulation possessing a thermal resistance of approximately 1·1 m² K/W, damage by cracking occurs – especially to the exterior walls. The continuous roof membranes are not divided into smaller spans by expansion joints, nor do they allow for the possibility for movement at their structural bearing.

In roofs where the thermal insulation layer is thin or where it is located below the slab, damage occurs in structural members longer than 10 m.

These cracks result partly from soaking of the brickwork by the penetration of rainwater. Lasting improvements frequently prove difficult as the cracks keep reappearing.

Points for consideration

— Rigid roof coverings often undergo changes of length as a result of drying-out (contraction) and temperature stresses. If these movements are restricted, stresses occur which can lead to failure (cracks) in the roof slab and also in the structural supports. Experience shows that linear expansion of up to 2 mm over both supports together produces no damage. Changes in length of this order of magnitude occur – according to the type of thermal insulation, the positioning of the layers and their fixing, production temperature and shape of the structure – in rigid roofing with a length of approx. 10–12 m.

— The magnitude of the temperature stress can be reduced by the provision of protective lamination above the thermal insulation (reduction of the summer surface temperature) and by the provision of exclusively external thermal insulating layers with a high thermal insulation value.

— The magnitude of the linear movement can be lessened by a reduction in span of the interacting structural members, in which the structural member is divided by an expansion joint and if possible has a fixed point in the centre.

— The damaging effect of linear movement can be prevented by free movement of the roof at the expansion joint and at the supports.

Recommendations for the avoidance of defects

● A rigid roof must have external thermal insulation with a thermal resistance of ⩾ 1·3 m² K/W.

● A rigid roof over 12 m in length must have expansion joints ⩾ 20 mm wide, unless the anticipated linear movement is proved by calculations to be harmless (see CP 110: Pt. 1: App. B) or the design of the supports allows for free bearing movement of the roof slabs.

(For further details on expansion joints and support design see A 2.3 – Bearing surface and expansion joints.)

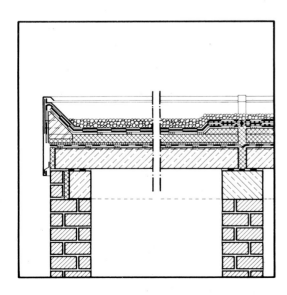

Solid slab flat roof
Sequence of layers and individual layers

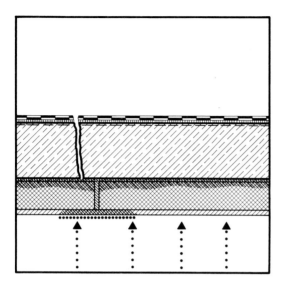

Half of all the damage caused by saturation on the typical cross-section occurs on roofs with a defective sequence of layers. Sometimes these are in roof constructions whose thermal insulation layer is on the underside of the roof slab or which, in addition to the upper thermal insulation, have a further insulating layer in the vapour barrier in the form of, for instance, suspended ceilings, boardings, acoustic slabs, extra internal insulation or lightweight pumice slag ceiling finish. Sometimes these are in roof constructions whose insulation to the interior is not protected by a vapour barrier. The condensation stress caused by this defect in the sequence of layers may be aggravated by rain penetration through leaks. In some cases – especially where there is great humidity inside – the damage from saturation with ensuing peeling of plaster and rejection of the sealing layer is caused solely by defective sequence of layers.

The roof slabs illustrated on the typical cross-section with internal insulating layers sometimes show cracks in the panel or bearing area.

Points for consideration

— For the correct functioning of the solid slab flat roof, free from condensation and thermally caused deflections of the rigid roof laminates, the sequence of layers, above all the correct positioning of the thermal insulation and the waterproof layers, is very important.

— Cross-sections of roof which show additional internal layers acting as thermal insulators, as well as thermal insulations without vapour barriers on the inside, permit the formation of water vapour and thus are liable to dampness from condensation.

— Internal thermal insulation layers do not provide the roofing slabs with protection from variations in temperature externally, but they reduce the protection effect of further external insulating layers. The result is increased thermally induced deformations which, if there is not sufficient possibility of movement, can lead to cracks in the roof slabs and the supports.

Recommendations for the avoidance of defects

● Where a solid slab flat roof has an external vapour barrier, this must have a vapour permeance value (diffusion resistance equivalent air layer thickness) of ⩾ 100 m and must be placed underneath the thermal insulating layers.

● All layers acting as thermal insulation should always be placed above the supporting roof structure. If this is not possible then either the thermal insulation above the vapour barrier must be strengthened or the anticipated increased thermally induced movements of the roof slabs must be taken into account in the construction of the expansion joints and roof supports.

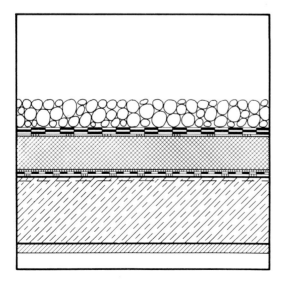

Solid slab flat roof
Sequence of layers and individual layers

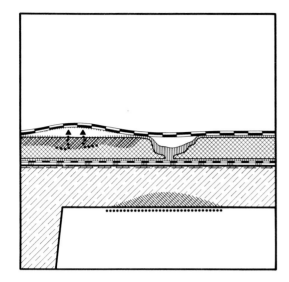

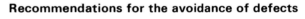

The formation of surface bubbles, damp thermal insulation slabs and sometimes the appearance of decay in thermal insulation made of vegetable-based material, indicate either that the insulation material was installed in a damp condition or that rainwater was trapped under the sealing layer while it was being laid. If insulation layers with open expansion joints are displaced or disturbed during adhesion, these defects will appear on the under-side of the roof as damp patches.

Points for consideration

– Dampness trapped between the vapour barrier and sealing layer – especially through insufficient protection of the thermal insulation material against damp during storing or construction of the roof – increases the danger of bubbles forming, and, with the use of vegetable-based materials, of rotting.

– Damaged insulation slabs, polystyrene foam panels destroyed by an adhesive which is too hot and panels which are not firmly fixed lead to thermal cold bridges which reduce the total thermal value and, above all, in lightweight roofs, can result in surface condensation on the under-side of the roof.

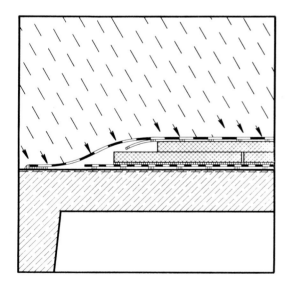

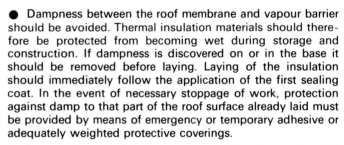

Recommendations for the avoidance of defects

● Dampness between the roof membrane and vapour barrier should be avoided. Thermal insulation materials should therefore be protected from becoming wet during storage and construction. If dampness is discovered on or in the base it should be removed before laying. Laying of the insulation should immediately follow the application of the first sealing coat. In the event of necessary stoppage of work, protection against damp to that part of the roof surface already laid must be provided by means of emergency or temporary adhesive or adequately weighted protective coverings.

● In order to avoid defective areas, insulation layers should if possible be of double thickness with concealed joints or, if single, should have a specified minimum overlap.

● In the selection of the adhesive or insulation material, care should be taken that a safe adhesion is achieved. However, a thermally or chemically caused deterioration of the insulating layer must be avoided.

Solid slab flat roof
Sequence of layers and individual layers

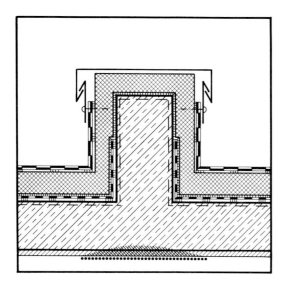

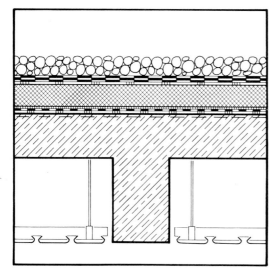

Where roofs have defective or badly calculated thermal insulation, large areas of damp are formed, especially on the underside of the ceiling below, where a change in use of the room will result in increased humidity.

In the case of suspended ceilings and of roofs with reinforced concrete upstand beams, the reinforced concrete roof beams show as dark stains on the ceiling. The solid roof slabs and reinforced concrete beams of such structures also show cracks.

Points for consideration

- Flat roofs are exposed to particularly strong thermal stresses. In the case of an unprotected black roof membrane, an exposed surface temperature difference of up to 100°C can occur. The thermal protection of flat roofs must therefore satisfy particularly high demands.

- Inadequately calculated thermal insulation gives poor thermal protection to the interior as well as to the structural parts of the building and results in such low internal surface temperatures that surface condensation can occur.

- The roof beams in hollow roofs which have an air space between the structure and ceiling finish (on account of their higher conduction of heat to the cavities) and roof coverings (on account of their large external cooling surfaces) become thermal bridges which on the inside allow the surface temperature to fall in places and lead to the formation of condensation.

Recommendations for the avoidance of defects

● Thermal insulation layers must be calculated so that the total cross-section of the structural member has a resistance to heat of $\geqslant 1\cdot 3$ m² K/W.

● In the case of solid roof slabs the thermal insulation resistance should be increased above the resistance of $1\cdot 3$ m² K/W in relation to the interacting length of the structural member, so that linear movements can be easily absorbed without damage (see A 1.1.3 – Linear expansion of roof slab).

● In the case of hollow roofs and similar roofs with heat-conducting support beams, the minimum transmitting resistance must be maintained at $0\cdot 8$ m² K/W for these unfavourable (as regards thermal insulation) areas of the cross-section.

● In principle, coatings should be avoided on solid slab flat roofs, as the thermal bridges thereby produced can be only unsatisfactorily obviated by increased thermal protection.

Solid slab flat roof
Sequence of layers and individual layers

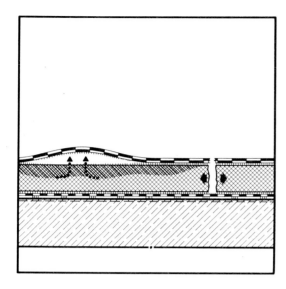

The defect of spot or zonal adhesion leads, in the case of sealing layers, to the formation of cracks, especially over the slab joints of the structural roof. The base support of the sealing coat consists partly of polystyrene foam slabs and glass-fibre whose form is not stable, of prefabricated composite slabs with metal foil lining and of aerated lightweight cement slabs.

The trapping of moisture and air in the area of thermal insulation between the under-side of the moisture barrier and the top of the sealing layer leads increasingly (as shown in the cross-section with no roof venting under the roof membrane) to the formation of bubbles.

No damage is caused by lack of a 'stabilising layer' under the moisture barrier.

Points for consideration

— Sealing panels that adhere over the whole of their base surface (e.g. thermal insulation) are totally subjected to the movements of the roof structure and can therefore become over-tensioned. These movements are especially noticeable on large-size bases constructed of aerated concrete slabs.

— Spot or strip adhesion reduces this tension, as the linear movement is taken up by a larger proportion of the surface.

— The reduced weatherproofing which this entails can be counterbalanced by an applied load of gravel ($\geqslant 50$ mm thick, of coarseness 16–32 mm); (see A 1.1.11 – Climatic demands on the sealing layer – Upper surface protection).

— Trapping of moisture and air under the roof membrane leads, if there is a high rise in temperature, to excess pressure and thence to the formation of bubbles. Venting under the roof membrane can compensate for this excess pressure. Because of high resistance to friction the effectiveness is, however, very limited and largely unrelated to whether the vents are connected at the edge of the structural member or by means of numerous roof vents on the surface of the roof.

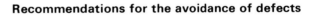

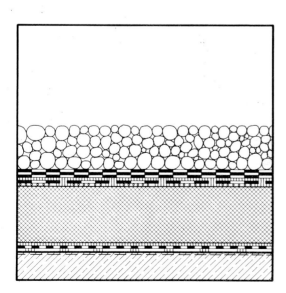

Recommendations for the avoidance of defects

● Between the roof membrane and the thermal insulation there should be a channel formed by a stabilising layer secured by spot or strip adhesion – e.g. made of glass-fibre bituminous roof sheets roughly sanded underneath, or of thermal insulation with closely spaced moulded channels on the upper side and self-adhesive roof sheet cover over.

● In the case of a large roof area made of aerated concrete roof slabs, in addition to spot adhesion the roofing panels over the expansion joints in the slab (transverse joints) must remain free for a width of 300 mm. This can be achieved by means of a loose-lying strip.

● The nature of the roof structure must permit a methodical laying of the sealing panels. It must also be flat, dry but not too porous, and have no oil or paint residue.

● Contact between the upper stabilising layer and the outside air is not necessary; it should indeed be avoided if the contact would be unusually complicated or the sealing layer perforated several times.

Solid slab flat roof
Sequence of layers and individual layers

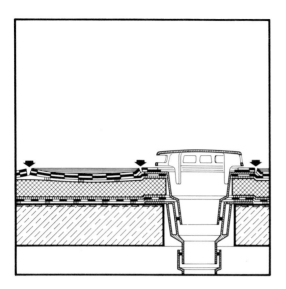

Leaks in the roof surface or at gutter outlets allow serious moisture penetration, especially on roofs with no fall.

Ponding on the upper surface of flat roofs leads to the formation of cracks in the roof membrane at the edge of the puddles, thus producing dampness. The same effect can be expected where puddles form around roof outlets which are too high, and also in channelled depressions with standing water. If a lightweight concrete screed has been used to produce the fall and has been applied directly to the roof structure under the moisture barrier, then moisture penetration quickly occurs. Sloping roof finishes above the thermal insulation will crack if the sealing layer is destroyed.

Points for consideration

– In roofs without falls there is constantly standing water, as the rain cannot flow away fast enough; this added weight of water increases the penetration possible through small fissures in the membrane.

– A roof cannot be constructed completely flat. Either the structural slab itself shows depressions caused in its construction or by deformation, or the roof cover produces unevennesses through overlapping. Therefore, in a roof without falls, ponding is inevitable.

– In the area of adhesion between the flange of the roof outlets and the main surface of the roof, the structural depth of the roof membrane is greater. If this increased depth is not taken into account, then the outlet lip lies above the level of the roof and this results in ponding.

– The area covered with water and the dry area of an unprotected roof membrane are simultaneously exposed to various stresses. At the edges of puddles there are irregular stresses which, together with the uneven build-up of dry dirt deposits, lead to cracks. In the area of constant standing water these cracks have serious consequences.

– Light concrete finishes laid to falls under the moisture barrier act as additional internal thermal insulation and therefore may give rise to condensation (see A 1.1.4 – Position of thermal insulation and vapour barrier in the cross-section).

Recommendations for the avoidance of defects

● The sealing layer should have ⩾ 3% (approx. 1·4°) fall to the outlets, especially where no upper surface protective layer is provided to combat dampness.

● The greater thickness of the roof in the area of the outlet flange and subsequent covering layers to the outlet should be compensated by a depression in the supporting structure.

● The necessary fall should be achieved by constructing the supporting slab with an inbuilt fall or providing a fall in the screed or insulating material. If the fall is formed by the thermal insulation then at the lowest point a heat resistance of ⩾ 1·3 m² K/W should be maintained.

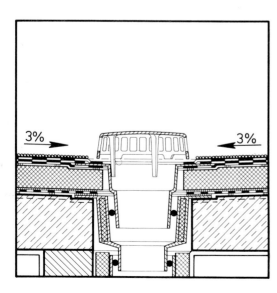

Solid slab flat roof
Sequence of layers and individual layers

In the main, sealing damage is due to leakages as a result of incorrect installation of the sealing layer. The defects occur mainly with multi-layer bituminous sheets, and on occasions with synthetic foils. Sometimes the overlap joint has become loose as a result of insufficient width of overlap or defective adhesion; sometimes there is inadequate adhesion of the sheets to each other or to the base.

If dirt, water or air is trapped during laying, then bubbles and folds appear on the surface of the sealing layer.

Points for consideration

– In a flat roof, especially one with no fall, careful formation of the sealing layer is of great importance.

– The waterproofness of sheet roofing depends largely on the careful formation of the overlap joint and, in the case of multi-layer roof membranes, on the complete adhesion of one to another over the entire surface.

– The trapping of impurities, water and air prevents complete adhesion of the layers and above all can lead to leakages in exposed surfaces unprotected from the sun's rays, due to climatic variation.

– Unprotected sealing layers with metal foil linings suffer considerable thermal expansion which can lead to loosening of the adhesive at the overlap joints.

– Wet weather and external temperatures $< +5°C$ prevent the setting of adhesion to the sealing layer.

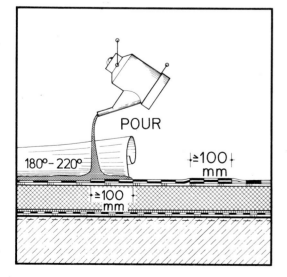

Recommendations for the avoidance of defects

● Sealing work on a flat roof should be carried out with special care and under strict supervision.

● Where sealing layers are made of bituminous roofing felt the longitudinal and transverse overlap seams should be ⩾ 100 mm wide. The sheets should be affixed together over their entire surface. The pouring and unrolling technique is preferable to other forms of adhesion, as this process offers the greatest guarantee of a vacuum-free and full surface adhesion.

● Heat sealing – at 180–220°C according to the type of bitumen – should be carried out only in dry weather and at outside temperatures of ⩾ +5°C.

● Roofing panels with metal linings may be used only in areas of low variation in temperature, e.g. under gravel fill and terrace finishes.

● Single-layer loose-laid synthetic roofing panels should receive added protection from damage by means of cover plates and warm foils, in addition to being glued or welded at their overlapping seams.

Solid slab flat roof
Sequence of layers and individual layers

Over 20% of damage by rainwater penetration in roofs as a result of fracture in the roof membrane can be traced back to mechanical damage to the sealing layer: this is most common during the period of construction.

The finished roof surface is commonly used as a working floor for carrying out other construction work – chimney-stack brickwork, etc. – in which heavy point loading occurs.

Bituminous as well as single-layer synthetic sheets are damaged by:

(i) excessive point loading (e.g. pressure by scaffolding trestles);

(ii) excess loading of lightweight insulation sheets (e.g. by people walking on them);

(iii) work on the (unprotected) sealing layer;

(iv) dropping of sharp-edged objects (tools and construction materials);

(v) walking on the gravel without support planks.

Points for consideration

– Sealing layers made of bituminous sheets, especially single-layer synthetic foils, are particularly vulnerable to mechanical stress. This problem increases if the sealing base consists of lightweight flexible thermal insulation material or if, under the sealing layer, there are hollows caused by open expansion joints or unevennesses.

Recommendations for the avoidance of defects

● Protective layers finished with a mineral coating, or preferably gravel fill, lessen the risk of damage. Necessary work on upstands where the sealing layer is often unprotected can be carried out only if appropriate precautions are taken: plank or board supports prevent pressure points made by trestles and damage through movement of gravel. Special care has to be taken in walking over the roof surface, especially at the angle between the upstand wall and the roof and where lightweight flexible insulating material is used as a base. There is often an open joint at this junction in the roof slab.

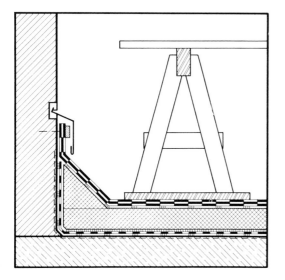

Solid slab flat roof
Sequence of layers and individual layers

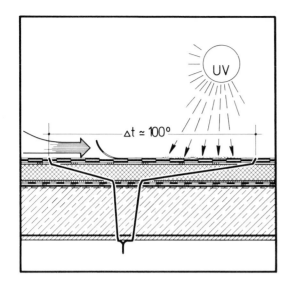

Damage by water seepage and the formation of bubbles and ponding are a result of climatic influences, occurring especially on unprotected or insufficiently protected sealing layers. The roof membrane when unprotected reveals various defects (see A 1.1.10 – Stressing of sealing layer).

The application of an upper surface protective sheet to reduce these stresses has been found to have its own problems:
(i) parts tend to be blown away by wind;
(ii) the thin reflective surface protective layer becomes dirty after a short time so that the roof membrane is exposed to stresses similar to those of the unprotected sealing layer.

Points for consideration

– In unprotected sealing layers differences in temperature of up to 100°C can occur. Large linear expansions and tensions, which can continue through into the structure below, have to be contended with.

– Unprotected sealing layers gradually become brittle under direct sunlight and the related ultra-violet rays, with resulting cracking and water penetration. With unprotected sealing layers the various tensions at the edge of the ponding are particularly high (see A 1.1.8 – Fall and formation of ponding on the roof membrane).

– There is considerable danger that unprotected sealing panels will be blown away, especially when there is no proper adhesion, nailing or bracing either to each other or to the base (see A 1.1.7 – Adhesion of roof membrane to base and A 1.1.9 – Laying of sealing layer).

– Roofs which are not designed for traffic are trodden on during construction work and the unprotected roof membranes are therefore liable to damage by mechanical influence (see A 1.1.10 – Stressing of sealing layer).

– Upper surface protective layers reduce these stresses on the roof membrane and counterbalance them.

– Upper surface protective layers can only function if equal distribution over the whole roof is guaranteed; if loose gravel is used, a coarseness of $\geqslant 16$ mm dia is needed to obviate the risk of its being blown away or collecting at water outlets.

– Upper surface protection comprising thin reflective layers loses its protective properties after a short time owing to weathering effects and dust deposits.

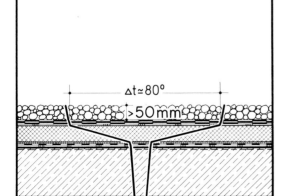

Recommendations for the avoidance of defects

● Basically, every flat roof should have an upper surface protective layer in the form of a loose gravel fill.

● A gravel fill applicable to a roof fall of 5° should be $\geqslant 50$ mm deep, having a coarseness of 16–32 mm dia (solid globule), and should be laid loosely and equally distributed over the entire roof. The weight of the gravel must be taken into account in the calculation and selection of materials for the loadbearing roof.

● Thin upper surface protective layers (reflective strips, sandings, etc.) must be regularly inspected at short intervals with regard to their correct functioning and, if found defective, renewed.

Solid slab flat roof
Typical cross-section

General texts and principles

Caemmerer, Winfrid: Wärmeschutz – aber richtig. Deutsches Bauzentrum, Köln 1958.

Cammerer, W. F.: Wärme und Feuchtigkeitsschutz. 15 Jahre Forschung. Erich Schmidt Verlag, Berlin 1969.

Eichler, F.: Bauphysikalische Entwurfslehre. Band 2, 4. Auflage. Verlagsgesellschaft Rudolf Müller, Köln 1973.

Gösele, K.; Schüle, H.: Schall Wärme Feuchtigkeit. 2. Auflage. Bauverlag GmbH, Wiesbaden und Berlin 1972.

Hoch, Eberhard: Flachdächer Flachdachschäden. Verlagsgesellschaft Rudolf Müller, Köln 1973.

Hoch, Eberhard: Kommentar Flachdächer. Verlagsgesellschaft Rudolf Müller, Köln 1971.

Jungnickel, Heinz u. a.: Abdichtungs- und Bedachungstechnik mit Kunststoffbahnen. Verlagsgesellschaft Rudolf Müller, Köln 1969.

Moritz, Karl: Flachdachhandbuch – Flache und flachgeneigte Dächer. 4. Auflage. Bauverlag, Wiesbaden und Berlin 1975.

Rick, Anton W.: Das flache Dach. 5. Auflage. Straßenbau, Chemie und Technik Verlagsgesellschaft mbH, Heidelberg 1966.

Seiffert, Karl: Wasserdampfdiffussion im Bauwesen. 2. Auflage. Bauverlag GmbH, Wiesbaden und Berlin.

Zentralverband des Dachdeckerhandwerks: Richtlinien für die Ausführung von Flachdächern. Ausgabe Januar 1973. Helmut Gros Fachverlag, Berlin 1973.

DIN 4108: Wärmeschutz im Hochbau, Ausgabe 1969 mit den ergänzenden Bestimmungen, Okt. 1974.

DIN 18338: Dachdeckungs- und Dachabdichtungsarbeiten, August 1974.

DIN 18530 (Vornorm): Massive Deckenkonstruktionen für Dächer, Dezember 1974.

Roof slabs

Balkowski, F. D.: Die Rißbildung am Deckenauflager. In: Das Dachdeckerhandwerk (DDH), Heft 2/75, Seite 88-91.

Brandes, Klaus: Dächer mit massiven Deckenkonstruktionen – Ursache für das Auftreten von Schäden und deren Verhinderung. In: Berichte aus der Bauforschung, Heft 87, Verlag Ernst & Sohn, Berlin 1973.

Braun, Hans: Aus der Praxis: Das gefällelose Dach als »Wanne«. In: Bauwelt, Heft 7/68, Seite 180–182.

Buch, W.: Temperaturmessungen an flachen Dächern. In: Bitumen, Teere, Asphalte, Peche..., Heft 10/73, Seite 405–409.

Grunau, Edv. B.: Flachdächer. In: Das Baugewerbe, Heft 15/74, Seite 19–24.

Hoch, Eberhard: Tragende Dachschalen aus Stahlbeton. In: Das Baugewerbe, Heft 7/74, Seite 30–34.

Künzel, H.; Gertis, K.: Untersuchungen über die thermische Beanspruchung von Gasbeton-Flachdächern. In: Deutsches Dachdeckerhandwerk (DDH), Heft 24/69, Seite 1518–1522.

Moritz, Karl: Praktische Erfahrungen und Details beim Flachdach. In: Das Dachdeckerhandwerk (DDH), Heft 18/70, Seite 1240–1256.

Pfefferkorn, Werner: Konstruktive Planungsgrundsätze für Dachdecken und ihre Unterkonstruktionen. In: Das Baugewerbe, Heft 18/73, Seite 57–65; Heft 19/73, Seite 54–59, Heft 20/73, Seite 86–90, Heft 21/73, Seite 54–63.

Rick, Anton W.: Fugenteilungen in Flachdächern. In: Deutsches Dachdeckerhandwerk (DDH), Heft 20/69, Seite 1338–1340.

Soyeaux, H.: Fugenausbildungen unter Verwendung von Polyisobutylen. In: Das Dachdeckerhandwerk (DDH), Heft 18/72, Seite 1312–1319.

Wolf, G.: Typische Schadensursachen bei Flachdächern mit Unterkonstruktionen aus profilierten Stahlblechen. In: Das Dachdeckerhandwerk (DDH), Heft 9/73, Seite 552–555.

Wolf, G.: Profilierte Blechtafeln und -bänder als Unterkonstruktion für Dachabdichtungen. In: Das Dachdeckerhandwerk (DDH), Heft 1/73, Seite 24–26.

Zimmermann, Günter: Volumenänderungen von Bauteilen. In: Deutsche Bauzeitung (db), Heft 3/69, Seite 188–196; Heft 5/69, Seite 350–364; Heft 7/69, Seite 524–526; Heft 11/69, Seite 844–848.

Vapour barrier – Thermal insulation

Balkowski, Dieter: Dachdecken mit zusätzlicher Wärmedämmung. In: Deutsches Dachdeckerhandwerk (DDH), Heft 12/66, Seite 636–637.

Buch, Werner: Das schwere und das leichte Flachdach. In: Deutsches Dachdeckerhandwerk (DDH), Heft 19/66, Seite 1029–1035.

Buch, Werner: Grundlegende bauphysikalische Fragen des Flachdaches unter besonderer Berücksichtigung des Wasserdampfproblems. In: Bitumen, Teere, Asphalte, Peche..., Heft 10/71, Seite 393–401.

Buch, Werner: Temperaturmessungen an flachen Dächern. In: Bitumen, Teere, Asphalte, Peche..., Heft 10/73, Seite 405–409.

Caemmerer, W.: Berechnung der Wasserdampfdurchlässigkeit und Bemessung des Feuchtigkeitsschutzes von Bauteilen. In: Berichte aus der Bauforschung, Heft 51, Verlag Ernst & Sohn, Berlin 1968.

Cammerer, W. F.: Der Wärme- und Diffusionsschutz des Flachdachs nach dem heutigen Forschungsstand. In: Bitumen, Teere, Asphalte, Peche..., Heft 10/71, Seite 402–412.

Eichler, Friedrich: Die »umgekehrte Dachdeckung« aus bauphysikalischer Sicht; aus: Wärme, Kälte, Schall, Heft 4, 1973.

Engels, Friedhelm: Flachdachisolierungen unter Verwendung fabrikmäßig vorgefertigter Bauelemente. In: Deutsches Dachdeckerhandwerk (DDH), Heft 89/68, Seite 1016–1017.

Gäbges, Peter: Die Dampfsperre im Flachdach. In: Bitumen, Teere, Asphalte, Peche..., Heft 10/70, Seite 435–440.

Götze, Heinz: Dämmschichten für Dächer und Außenwände. In: Das Bauzentrum, Heft 3/72, Seite 65–85.

Grün, Wolfgang; Müller, H. J.: Sind Dampfsperren im Warmdach wirklich fragwürdig? Der Dachdeckermeister, Heft 10/70.

Haefner, R.: Bauschäden an Flachdächern. In: Industrie-Anzeiger, Heft 42/68, Seite 850–854.

Haushofer, Bert; Wichmann, H.: Vollwärmeschutz aus der Sicht des Dachdeckers. In: Das Dachdeckerhandwerk (DDH) Heft 19/74, Seite 1256–1262.

Haushofer, Bert: 98 Prozent Luft – 16 Jahre Erfahrung, Langzeiterfahrungen mit Styropor auf dem Flachdach. In: Deutsches Dachdeckerhandwerk (DDH), Heft 1/68, Seite 702–708.

Haushofer, Bert: Möglichkeiten und Problematik des Bauelementes für das Flachdach – Dach-Konstruktion. In: Das Dachdeckerhandwerk (DDH), Heft 4/70, Seite 195–198.

Hebgen, Heinrich; Heck, Friedrich: Einschalige Flachdächer mit verschiedenen Dämmstoffen. In: Deutsches Dachdeckerhandwerk (DDH), Heft 16/68, Seite 988–992.

Heck, Friedrich: Kältebrücken in der Dämmschicht beim Flachdach. In: Deutsches Dachdeckerhandwerk (DDH), Heft 16/68, Seite 1008–1011.

Hoch, Eberhard: Gasbetondecken im Flachdachbau. In: Das Dachdeckerhandwerk (DDH), Heft 9/73, Seite 550–552.

Niehoff, Hans: Dachdämmplatten aus Polyurethan-Hartschaumstoff. In: Das Dachdeckerhandwerk (DDH), Heft 19/72, Seite 1390–1395.

Noeße, Rudolf: Zweckmäßiger Aufbau von Flachdächern aus Gasbeton. In: Das Dachdeckerhandwerk (DDH), Heft 15/70, Seite 1034–1037.

Probst, Raimund: Dachdecken. In: Das Bauzentrum, Heft 4/68, Seite 13–17.

Probst, Raimund: Dächer- »kalt« oder »warm«. In: Baupraxis, Heft 5/69, Seite 52–56.

Probst, Raimund: Bauschaden des Monats März/Juni. In: Deutsche Bauzeitung (db), Heft 4/72, Seite 310–312.

Rick, Anton W.: Möglichkeiten der Nachdämmung bei einschaligen Flachdächern. In: Das Dachdeckerhandwerk (DDH), Heft 2/72, Seite 109–110.

Rick, Anton W.: Sind Dampfsperren wirklich unnötig? In: Deutsches Dachdeckerhandwerk (DDH), Heft 14/69, Seite 872–873.

Rick, Anton, W.: Grenzen der Wärmedämmung von Flachdächern. In: Bitumen, Teere, Asphalte, Peche..., Heft 25/74, Seite 360–363.

Riedel, Peter: Schaumglasdämmungen ohne Dampfsperre. In: Deutsches Dachdeckerhandwerk (DDH), Heft 2/67, Seite 83–85.

Schild, Erich: Planung, Ausschreibung und Ausführung von Warmdächern (Erkenntnisstand 1972). In: Deutsches Architektenblatt (DAB), Heft 11/72, Seite 733–736.

Schüle, Walter; Jenisch, Richard: Kondensations- und Austrocknungsverhältnisse bei nichtbelüfteten Flachdächern. In: Berichte aus der Bauforschung, Heft 80, Verlag Ernst & Sohn, Berlin 1973.

Schumacher, R. v.: Extrudierte Polystyrol-Schaumstoffe für den Dachdecker. In: Deutsches Dachdeckerhandwerk (DDH), Heft 21/69, Seite 1406–1412.

Seiffert, Karl: Gibt grundsätzliche Vorteile beim einschaligen oder beim zweischaligen Dach? In: Bitumen, Teere, Asphalte, Peche..., Heft 18/67, Seite 370–374.

Falls – Sealing layer – Upper surface protection

Andernach, W.: Die Verklebung von Dachbahnen. In: Bitumen, Teere, Asphalte, Peche..., Heft 10/66, Seite 378–380.

Brocher, Erichbernd: Schäden an Flachdächern durch Luftstauatmung. Aus: »wirtschaftlich bauen«, Heft 2/68, Seite 67–69.

Buch, Werner: Temperaturmessungen an flachen Dächern. In: Bitumen, Teere, Asphalte, Peche..., Heft 10/1973, Seite 405–409.

Gründler, Klaus: Wärmedämmung und Gefälleschicht bei einschaligen Flachdächern. In: Das Dachdeckerhandwerk (DDH), Heft 10/70, Seite 646–647.

Gockel, Franz: Dampfdruckausgleich und Dampfsperre beim Flachdach. In: Deutsches Dachdeckerhandwerk (DDH), Heft 6/65, Seite 240–242.

Götze, Heinz: Die lose verlegte Kunststoff-Dachhaut. In: »Kunststoffe im Bau«, Themenheft 27.

Haage/Kramer: Neue Erkenntnisse über die Windbelastung auf Flachdächern. In: Das Dachdeckerhandwerk (DDH), Heft 14/74, Seite 1446–1448.

Halász, Robert von: Neue Erkenntnisse und praktische Folgerungen für die Ausbildung von Flachdächern. In: Bitumen, Teere, Asphalte, Peche..., Heft 5/63, Seite 218–220.

Haushofer, Bert: Heißbitumenklebemassen aus der Sicht des Dachdeckermeisters. In: Deutsches Dachdeckerhandwerk (DDH), Heft 6/65, Seite 244–247.

Haushofer, Bert: Beitrag zur besseren Verklebung bituminöser Dachbahnen. In: Deutsches Dachdeckerhandwerk (DDH), Heft 16/67, Seite 892–893.

Haushofer, Bert: Bituminöse Schweißdachbahnen. In: Bitumen, Teere, Asphalte, Peche..., Heft 10/69, Seite 471–479.

Haushofer, Bert: Lagen, Leitern, Löcher! In: Das Dachdeckerhandwerk (DDH), Heft 1/72, Seite 22–25.

Haushofer, Bert: Untere Entspannungsschicht: Ja – nein – Ja! In: Deutsches Dachdeckerhandwerk (DDH), Heft 15/66, Seite 800–802.

Hoch, Eberhard: Ortschaumstoffdächer eine zukunftsweisende Entwicklung? In: Das Dachdeckerhandwerk (DDH), Heft 1/74, Seite 16–18.

Holzapfel, W.: Abdichtungsbahnen aus Kunststoff. In: Das Dachdeckerhandwerk (DDH), Heft 1/75, Seite 16–25.

Holzapfel, W.: Dachbahnen. In: Das Dachdeckerhandwerk (DDH), Heft 2/72, Seite 102–106.

Keller, von: Das flache Dach im Windangriff. In: Das Dachdeckerhandwerk (DDH), Heft 21/70, Seite 1490–1492.

Kulhavy, A.: Oberflächenschutz bei verschiedenen Dachneigungen. In: Deutsches Architektenblatt (DAB), Heft 17/74, Seite 1161–1162.

Müssig, Hans-Joachim; Hummel, Rudolf: Dachflächen grundsätzlich mit Gefälle ausbilden? Kommentierung der Flachdachrichtlinien. In: Das Dachdeckerhandwerk (DDH), Heft 1/74, Seite 53–54.

Osterritter, K.; Peter, H.: Zur Diskussion: Druckausgleichsschicht, Entspannungsschicht, Entfeuchtungsschicht, Überbrückungsschicht, Dampfsperre, Blasenbildung. In: Bitumen, Teere, Asphalte, Peche..., Heft 1/64, Seite 8–10.

Rick, Anton W.: Verlegen von Deckungen und Dichtungen auf feuchten Flächen bei kaltem Wetter. In: Deutsches Dachdeckerhandwerk (DDH), Heft 9/69, Seite 520–523.

Rick, Anton W.: Dachbahnen auf stärker geneigten Flächen. In: Deutsches Dachdeckerhandwerk (DDH), Heft 18/69, Seite 1172–1173.

Rick, Anton W.: Zur Größenordnung von Dampfwirkungen und zur Blasenbildung. In: Bitumen, Teere, Asphalte, Peche..., Heft 3/1967, Seite 93–99.

Rick, Anton W.: Kreuzverband und lange Bahnen? In: Das Dachdeckerhandwerk (DDH), Heft 1/72, Seite 36–37.

Rick, Anton W.: Die Wirkung der Wärmedämmung auf die Alterung von Dachpappen-Deckschichten unter dem Einfluß der Sonneneinstrahlung. In: Bitumen, Teere, Asphalte, Peche..., Heft 3/1965, Seite 122–124.

Rick, Anton W.: Warum Dachabkiesungen? In: Bitumen, Teere, Asphalte, Peche..., Heft 20/1969, Seite 473–474.

Rick, Anton W.: Ebenheit von Bauflächen. In: Bitumen, Teere, Asphalte, Peche..., Heft 10/73, Seite 426–428.

Rick, Anton W.: Zur Frage der ebenen Dächer. In: Bitumen, Teere, Asphalte, Peche..., Heft 3/66, Seite 112–114.

Schild, Erich: Bauschäden und Baufolgeschäden an Flachdächern. In: Industrie-Anzeiger, Heft 74/68, Seite 1666–1668.

Seiffert, Karl: Der Glaube an die Entspannungsschicht. In: Bitumen, Teere, Asphalte, Peche..., Heft 10/70, Seite 421–425.

Wald, Hans A., Kelcher, Hans: Folien aus Polyäthylen-Bitumen-Kombination für die Dachabdichtung. In: Bitumen, Teere, Asphalte, Peche..., Heft 2/72, Seite 63–78.

Werner, Hugo: Deckungen schwerer und leichter Flachdächer. In: Deutsches Dachdeckerhandwerk (DDH), Heft 15/66, Seite 781–784.

Zimmermann, G.: DAB-Bauschädensammlung 6.4.1974: Einschaliges Flachdach mit Bituminöser Dachhaut-Durchfeuchtungen der Dachkonstruktion infolge mechanischer Beschädigung der Dachhaut. In: Deutsches Architektenblatt (DAB), Heft 12/75, Seite 928.

N. N.: Flachdach-Dichtung und -Entwässerung. In: Bauen mit Kunststoffen, Heft 4/74, Seite 231–248.

Problem: Connection to vertical abutments

On most flat roofs the roof structure must be connected to an adjoining vertical structure. It may be to a chimney-stack, another roof structure, high parapets and railings, or to an existing higher adjacent building.

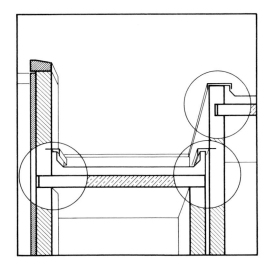

Empirical research investigations into structural defects have shown that this connection of the sealing layer and the adjoining vertical surface is a definite problem on solid slab flat roofs.

Of the number of roof details studied, more than one in five showed defects directly attributable to this problem. This shows the highest incidence of damage recorded for a single detail.

Analysis shows that this high incidence is due to certain typical, frequently recurring defects. If these defects are avoided in the design and construction of roofs, the risk of failure can be considerably reduced.

In the following examples of connections to abutting vertical walls the primary function of the detail, i.e. the creation of a watertight junction between the roof membrane and adjacent structural wall, was not achieved in the long term.

Solid slab flat roof

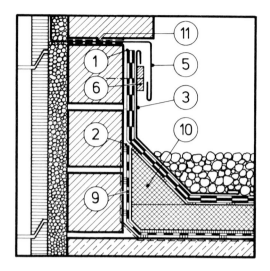

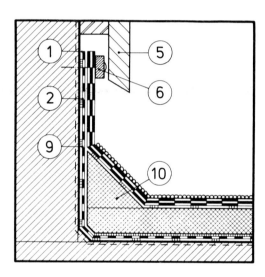

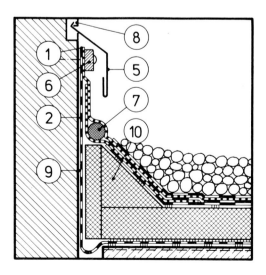

1 The roof membrane, at its connection to the vertical wall, must be raised above the highest level of water that may be expected in bad weather conditions. The fillet must be $\geqslant 150$ mm above the roof surface – gravel fill, compressed mineral layer or roof membrane (see A 2.1.2).

2 The edge of the fillet and the connection of the sealing layer to the surface of the vertical abutment must be formed either from the roof membrane itself or by using a connecting foil which allows for the extreme stresses at this junction. A frictional connection of the pliable roof membrane with metal profiles or plates should be avoided (see A 2.1.4).

3 The raised roof membrane must be protected near the edge of the fillet from extreme effects of external climate and from mechanical stresses (see A 2.1.4).

4 If the use of continuous pressed metal sheets directly fixed to the wall is unavoidable then these should be provided with movement joints at adequate intervals (approx. every 3 m) (see A 2.1.4).

5 The termination of the roof sealing layer must be protected from the force of water streaming down the face of the abutment wall (see A 2.1.5).

6 In order to avoid loosening or slipping of the vertical sealing layer, this should be fixed regularly along its whole upper length and protected by a metal cover flashing (see A 2.1.5).

7 If movements are expected between the vertical member and the sealing base – e.g. due to differential settlements – then the sealing layer should not be fixed to the vertical surface unless it is in a position to take up these deformations (see A 2.1.5).

8 Overhanging sheets intended to fulfil a flashing function (water diversion) should be installed according to instructions and should be subdivided by sliding seams (in the case of titanium zinc, for example, at maximum centres of 5–6 m) (see A 2.1.5).

9 The roof layer above the structural slab, particularly the thermal insulation layer, should be protected at the edge against possible damp penetration from above, e.g. by the adhesion of the vapour barrier (see A 2.1.3).

10 The transition between the flat roof surface and angle fillet of the roof should be as uniform as possible, avoiding sharp folds and bends to built-up layers in the area of the abutment.

11 For these details to function well, the connection between the roof's built-up sealing layers and its structural part is of paramount importance. A continuous impervious barrier must be provided for claddings or other non-waterproof facings.

Solid slab flat roof
Connection to vertical abutments

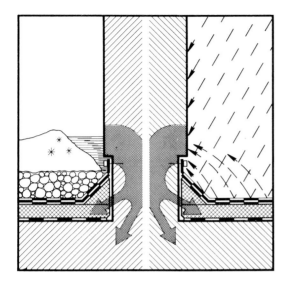

If the sealing layer is not placed at a sufficient height above the upper surface of the roof, damage by water penetration to the roof structure (thermal insulation layer) and internally to the rooms below will ensue. Where the roof surface has a shallow fall (< 6°) and at the lowest points of the roof, the raised angle of the sealing layer will be submerged as a result of backwash and the damming of water by wind and melting snow.

Similarly, the vertical abutments (e.g. walls) will be saturated and penetration to the surface of the inner wall by locally retained water or spray will occur.

Points for consideration

– Surfaces of vertical walls are subjected to unusual tensions from water and spray retained in the area of adjacent roof surfaces, similar to those occurring in the foundations to the ground floor. Here special measures to protect against damp are necessary.

– The termination of the angled sealing layer where it joins the vertical face cannot, in practice, be made permanently watertight.

Recommendations for the avoidance of defects

● At its junction with a structural wall the roof membrane must be placed sufficiently high above the level at which water would penetrate in the worst situation anticipated. The ultimate height of the sealing membrane must be ⩾ 150 mm above the upper surface of the roof — gravel fill, compressed gravel layer or roof skin. In special circumstances higher levels may be necessary, e.g. where channels are formed in the roof surface or with unavoidable flooding of a tank roof.

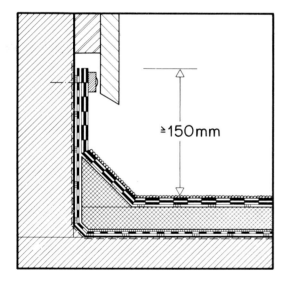

≧150mm

Solid slab flat roof
Connection to vertical abutments

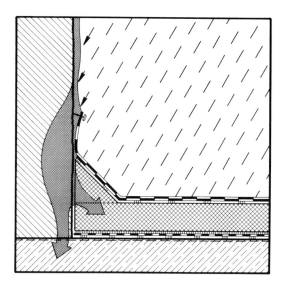

Rainwater may penetrate behind the edge of the raised sealing layer at its connection to the vertical surface (e.g. wall). This is the result of (a) inadequate protection to the ends of the sealing layer, which are fixed to the face of the abutting wall, or (b) vertical sealing layers which are inadequately fixed to the structural surface, being only spot fixed or glued on, and have come loose. Similar problems occur where metal flashings, provided to protect against damp, have worked loose from the joint or pulled apart at the soldered seam, leaving the edge of the sealing layer exposed to water.

Points for consideration

— The higher a vertical wall surface and the more freely it is exposed to driving rain, the heavier the water run-off from its surface. Every junction of the sealing layer which protrudes from this surface presents an obstacle to the running water and is strongly attacked by wind and water.

— The surface of the abutment wall is usually not completely uniform so that, for example, a pressed metal flashing finds no even and continuous bearing surface. Sealing by cement has very limited durability.

— The built-up roof membrane is subjected to strong thermal influences in the vertical plane, which can adversely affect its stability and that of the adhesion layer or its fixings and, as previously mentioned, pressed metal flashings are exposed to strong thermal deformations.

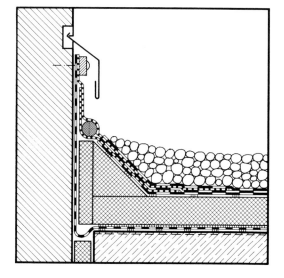

Recommendations for the avoidance of defects

● The edge of the sealing layer must be protected from penetration by running water at the junction of the wall and roof surfaces. This automatically occurs if the end of the seal is re-set behind the water-diverting upper surface or goes under the flashing.

● In order to avoid loosening or slipping of the vertical sealing layer, this should be fixed at its upper end with a clamp rail at regular intervals along its whole length.

● If movements are expected between the abutment wall and the sealing layer – e.g. due to variable settlements – then the sealing layer should not be fixed to the wall except by means of an expansion loop placed in a position to absorb the resulting deformations.

● Pressed metal or lead flashings which should fulfil a sealing function (water-diversion) must be able to move freely and must be provided with sliding seams at their joints.

● The roof member above the structural slab and, most important, the thermal insulating layer should be protected at the sides against possible water penetration from above, e.g. through the adhesion around the vapour barrier.

Solid slab flat roof
Connection to vertical abutments

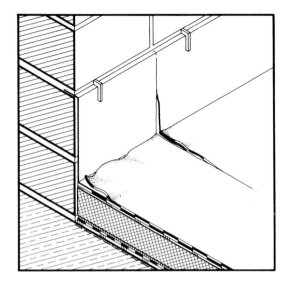

Where sheet metal or one-part metal profiles are used to form the junction of the sealing layer and the vertical structural wall, leakages are frequently observed. Long lengths of sheet metal without provision for expansion have been found to exhibit torn soldered seams and warping. If the metal adheres firmly to the roof skin, then this cracks, especially at the soldered seam. With unprotected areas of adhesion the built-up roof membrane separates from the adhesion surface.

Points for consideration

— The use of sheet-metal angles necessitates the permanent, watertight connection of two very different structural materials by means of an adhesive.

— The vertical faces of the sealing layer are normally exposed to climatic influences and particularly to changes in temperature. Depending on the material used, large linear expansion and changes of form can arise which stress excessively the roof skin or other frictionally connected structural parts, leading to failure by compression (warping, folds) and tension (cracks). These stresses can be reduced by dividing the effective expansion lengths by means of movement joints. However, such provision for expansion is expensive and may give rise to new defects; wherever possible, firmly adhering wall connection sheets should be avoided.

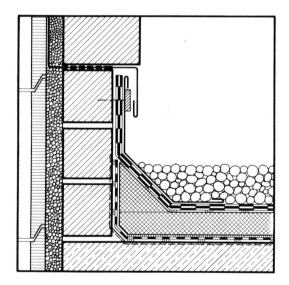

Recommendations for the avoidance of defects

● The fillet cover and the junction of the sealing layer to the vertical abutment wall should be formed from the roof membrane itself or by using connecting foils of similar material behaviour but better able to absorb extreme stresses. A frictional connection of the pliable roof membrane with metal profiles or sheet metal should be avoided.

● The vertical roof membrane must be protected from the extreme effects of the climate and from mechanical stresses in the area of the junction. This is achieved by the application of cover flashings made of sheet metal. Sheet-metal profiles used as flashings must therefore move freely (i.e. not be firmly built in) and must be adequately subdivided by expansion joints in the form of overlap joints (with titanium zinc sheet, $\geqslant$ 5–6 m).

● If the use of wall connection sheets is unavoidable, these should be provided with movement joints at adequate intervals (with zinc sheets, $\geqslant$ every 3 m). The metal sheet, after removal of damp and dirt, should be bonded with adhesive for a minimum width of 120 mm to the multi-layer roof membrane. On sloping and vertical surfaces, special adhesives must be used.

Solid slab flat roof
Connection to vertical abutments

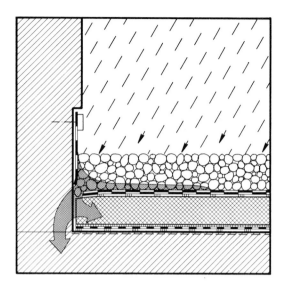

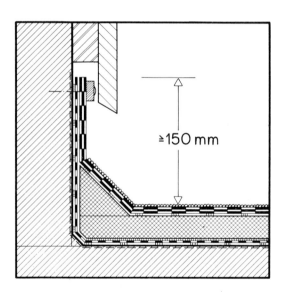

In the angle of a roof where the built-up membrane of soft connecting foils – e.g. synthetic foils – changes plane, penetration has been found to result from mechanical damage: the pliable vertical and horizontal layers deflect, forming a right angle, and water leaks in where there is inadequate support at the junction.

Points for consideration

– Soft sealing sheets and foils cannot be sharply angled, particularly in a built-up formation, without damage. In normal building practice, therefore, those in the junction between the wall and the roof must be supported.

– At hollow places, as well as where there is a flexible base, the sealing layer is particularly susceptible to mechanical damage.

Recommendations for the avoidance of defects

● The transition between the roof surface and the abutment wall should, if possible, be designed so that a sharp angle at their junction is avoided. A gradual transition must be allowed for in the sealing layer – e.g. by the use of triangular fillets made of insulating foam or impregnated timber sections – so that a continuous sealing base is maintained. At the junction of buildings, purpose-made fillets should be provided if necessary.

Solid slab flat roof
Connection to vertical abutments

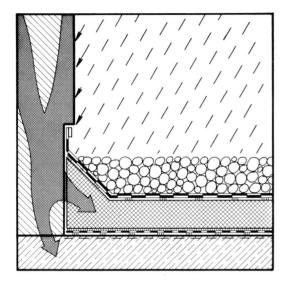

A normally acceptable junction detail of the roof membrane to the vertical surface of the wall will not in every case prevent the entry of water into the roof structure. Rainwater may find its way inside the wall and flow behind the sealing layer. This occurs mainly in chimney-stacks and incorrectly installed claddings to external walls.

Points for consideration

– Protection against damp to outside walls, especially those constructed of more than one material, is not necessarily confined to the outer surface of the wall; the whole thickness of the wall must receive protection.

– Chimney-stacks are subject to considerable changes in temperature and consequently to deformation which may lead to cracks in the chimney coping or in the outer wall. Simultaneously, they are exposed to weathering effects and the whole outer wall may be saturated.

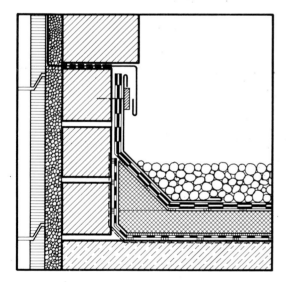

Recommendations for the avoidance of defects

● The correct functioning of the connection detail depends on the effectiveness of the connection of the roof membrane to the waterproof structural layer. Porous facings and claddings must be isolated from the sealing layer.

Problem: Edge of structural member

The majority of solid slab flat roofs are designed with internal drainage points, single outlets or drainage channels. These determine decisively the structural design of the perimeter of the roof. Rainwater must be channelled quickly away from the edge of the roof in order to prevent an overflow at the edge.

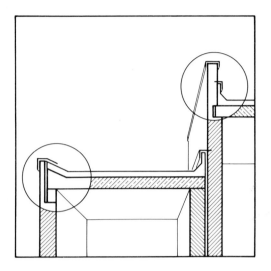

Damage to the edges of solid slab flat roofs with external drainage had not occurred among the examples investigated. Roofs with internal drainage points showed (in relation to the total number of instances of damage to points of detail of this type of roof) a large number of defects at the roof edge itself.

After the evaluation of a representative quantity of examples, typical and recurring faults in design or construction were recognised as causing problems. In many instances the necessary edge slope to the roof membrane had been omitted or was ineffective: protection to the termination of the sealing layer in particular was disregarded. The following pages deal with these defects in more detail and recommendations for their avoidance are proposed.

Solid slab flat roof

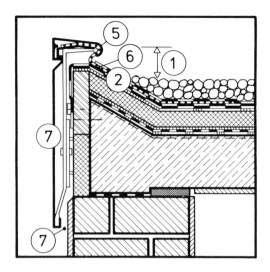

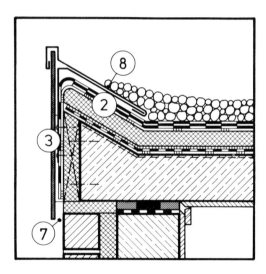

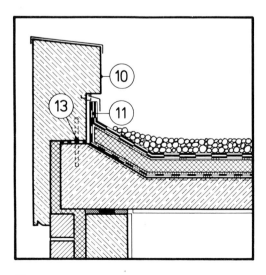

1 The edge of a roof drained inwardly must be raised above the drainage plane. The sealing layer must be raised ⩾ 100 mm above the upper surface of the roof – gravel fill, compressed gravel layer or other roof membrane (see A 2.2.2).

2 The roof edge should have an effective fall to the roof surface by means of a built-up angle of ⩾ 30° (see A 2.2.2).

3 At the edge, the built-up roof membrane should be guided over an angle preformed in the sealing base to the required height, externally carried over the edge and fixed by nailing or clamp rails to the vertical face of the roof behind the applied fascia (see A 2.2.3).

4 A frictional connection of the pliable roof membrane with cover trim made of metal, asbestos cement or synthetic material should be avoided (see A 2.2.5).

5 If the sealing layer is connected to a roof edge cover trim, then this must securely anchor the roof membrane; at the same time, however, transfer of the deflections of the edge trim to the fastened roof membrane must be avoided (see A 2.2.4).

6 The roof membrane must be laid on a secure angled slope until it is fastened in the edge cover trim (see A 2.2.4).

7 The fixing of the roof edge trim must be secured to the supporting members of the roof edge, so that when load is applied there is no deflection with culminating stress effect on the roof membrane and the fascia, which is at a distance of ⩽ 20 mm in front of the external wall surface (see A 2.2.4).

8 The tilted roof edge must be protected by a clipped-on pressed metal cover trim which allows movement and by loose fill gravel which is not connected with the sealing layer (see A 2.2.3).

9 High parapets at the roof edge should be avoided if possible (see A 2.2.6).

10 If parapets are provided, expansion joints must be arranged at centres of 5 m maximum. These expansion joints should be carefully sealed (see A 2.2.6).

11 The junction of the roof membrane with the parapet should be designed so that the anticipated movements of the parapet section – above all in the area of the joints – do not lead to damage to the sealing layer (see A 2.2.6).

12 Brickwork parapets should be protected (see A 2.2.6).

13 Parapet units which are positioned on the roof slabs, but with no direct adhesive contact, only anchored, are preferable to those fixed directly to the roof slabs or the external brick-work (see A 2.2.6).

Solid slab flat roof
Edge of structural member

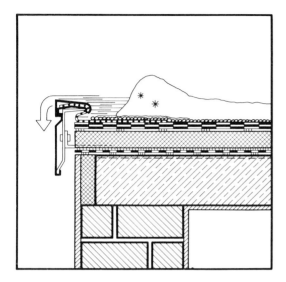

Failure has been repeatedly found to occur in roofs where the edge is not raised, or where it is only marginally raised above the main roof level. In roofs which have no fall or are designed with virtually no fall, the rainwater, because of defective falls or wind pressure, can overflow the edge and lead to heavy saturation of the outside walls. If the roof edge is inadequately sealed the roof structure is penetrated and becomes saturated.

This defect is easily recognisable in roofs which have interlocking metal overhanging sections. These sections sag and no longer drain towards the inner part of the roof so that, the edge slope being low, the rainwater runs over the roof edge (see A 1.1.2 – Deflection of loadbearing slab).

If the roof surface is covered with gravel granules < 16 mm dia this loose gravel layer will be blown by the wind over the roof edge. Such falling gravel has been known to cause damage.

Points for consideration

— The roof edge may be exposed to increased water pressure through a defective fall or a fall which is minimised due to sagging of parts of the roof. Excess water loading can also be generated by wind banking at the roof edge, or a backwash can be caused by blocked drainage outlets on the roof, resulting in increased water pressure. Flat roof edges without an upstand cannot retain this increased amount of water.

— Lightweight loose gravel granules can be blown by the wind; these can cause a build-up over part of the roof or they can be blown over roof edges.

Recommendations for the avoidance of defects

● The edges of roofs drained inwards must be raised above the level of the main flat roof to avoid the onrush of water which can be expected in unfavourable conditions. The sealing layer must then be ⩾ 100 mm above the upper surface of the roof – gravel fill, compressed gravel layer, or any other roof membrane. In special circumstances – at the lowest point on the roof – the upstand may have to be higher.

● The roof edge should have an effective angle of fall to the roof surface of ⩾ 30°.

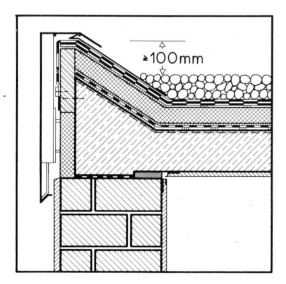

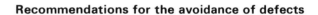

≥100mm

Solid slab flat roof
Edge of structural member

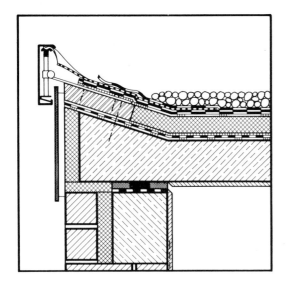

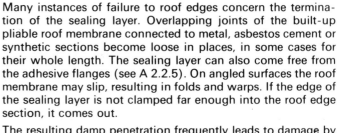

Many instances of failure to roof edges concern the termination of the sealing layer. Overlapping joints of the built-up pliable roof membrane connected to metal, asbestos cement or synthetic sections become loose in places, in some cases for their whole length. The sealing layer can also come free from the adhesive flanges (see A 2.2.5). On angled surfaces the roof membrane may slip, resulting in folds and warps. If the edge of the sealing layer is not clamped far enough into the roof edge section, it comes out.

The resulting damp penetration frequently leads to damage by saturation to large parts of the roof structure and the rooms below.

Points for consideration

— Connections and joints in the sealing layer in the proximity of the roof edge, which are exposed to climatic influences, are vulnerable because of the required bending of the layers and the easily damaged edges of the strips and foils. This is especially true of connections of the sealing layer to materials with different properties (e.g. metals).

— If the roof membrane ends at the edge or is clamped in trims, the weathering face of the roof structure is exposed during construction if there is no protecting fascia; (joints in sheet-metal cladding must be protected from behind) (see A 2.2.5).

Recommendations for the avoidance of defects

● At the roof edge the membrane should be laid over an angled upstand preformed in the roof slab to the required height of 100 mm; it should then be turned over the upstand and fastened on its face by means of nails or clamp rails.

● The angled roof edge should be protected by interlocked, in some cases pre-sprung, trims laid loosely on the sealing membrane and protecting the junction.

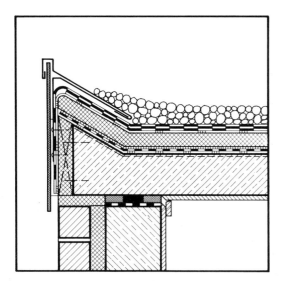

Solid slab flat roof
Edge of structural member

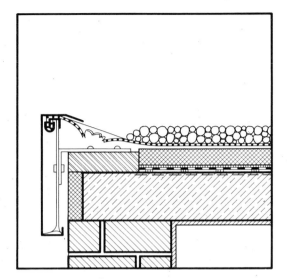

In roof designs with edge-connecting foils clamped into multi-part cover trims, if these are laid partly under stress or laid unsupported they will be damaged by traffic on the roof. If the multi-part cover trim is covered with adhesive, cracks occur at its junction with the sealing membrane. At roof edges where the sealing layer does not adhere to the trim and with fascias erected at a considerable distance from the wall surface, damage occurs in conditions of high wind velocity.

Points for consideration

— The strengthening of the edge of the sealing layer by metal cover trims requires a form of construction that anchors the roof membrane securely, but which avoids transference of the deformation of the roof edge sections to the roof membrane. This can be achieved by means of multi-part cover trim with clamping fixings.

— If the edge upstand is formed by means of the metal section alone, the roof membrane is unsupported. The stretched and unsupported sealing membrane is particularly vulnerable to mechanical stresses.

— Wind stress to the roof is increased if, added to the suction effect on the roof surface, there is a considerable space between the fascia and the wall. This open area allows the wind pressure to build up directly below the sealing membrane, which is unsupported, and causes it to crack.

— Contamination of the cleat section by adhesive, etc., as well as inaccurate erection, can lead to unintended frictional connections of the roof membrane and the cover trim and cause cracking.

— The cleats and cover trims are subjected to other external stresses under certain circumstances, e.g. leaning of ladders. These trims, if of slender section or if not fixed to the structural part of the roof edge, may be disfigured so that the roof membrane contained in the trim is unduly stressed.

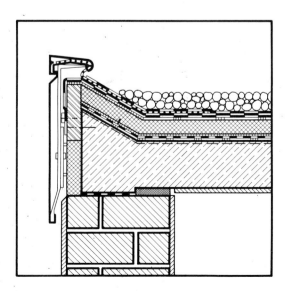

Recommendations for the avoidance of defects

● If the sealing layer is connected to a roof edge cover trim then this must anchor the roof membrane securely; however, the transfer of deformations of the edge trim to the roof membrane must be avoided under normal building conditions. This can be achieved by the use of a multi-part detachable cover trim.

● The pliable sealing roof membrane must be laid continuously over the supporting angled slope of the roof and into the edge of the cover trim.

● The securing of the roof cover trim to the loadbearing part of the roof edge must be such that, when stresses take effect, no deformation results that will affect the clamped roof membrane; the fascia should be at a distance of $\leqslant 20$ mm in front of the external wall surface.

Solid slab flat roof
Edge of structural member

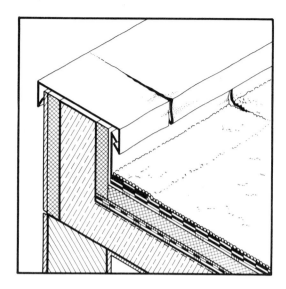

In roofs with edges formed of metal trims frictionally connected to the sealing membrane, or upstands covered with pressed metal sections, there are often leaks leading to the subsequent saturation of the roof structure and the rooms below.

The following specific defects have been observed.

(i) In a built-up roof membrane where the sealing layer adheres to the metal cover section, cracks appear in the roof membrane near the open joints or at the soldered seams of the metal trims.

(ii) The adhesive between the roof membrane and the metal surfaces deteriorates in places, sometimes along its whole length.

(iii) At inclined unprotected adhesion points, the sealing membrane separates from the sheet metal and leaks appear, particularly where the area of adhesion is too small.

(iv) In areas where large sheet-metal copings are used at the roof edge (e.g. parapet coping), warping, buckling and cracked soldered seams occur through which rainwater penetrates the roof structure and the external walls.

Points for consideration

— To seal the roof edge upstand successfully, employing sections made of metal, asbestos cement or synthetic material, requires the permanent watertight joining of two very different structural materials by means of an adhesive.

— The roof edge is exposed, without protection, to climatic effects and in particular to changes in temperature. According to the material used, considerable linear expansion and changes in form can be generated, which excessively stress the roof membrane or other frictionally connected components and lead to failure by warping and cracking. By subdivision of the effective linear expansion and provision of expansion joints, these deformations are reduced; however the construction of such provision for expansion is expensive and conceals new weaknesses.

— Strong sunlight on the often unprotected adhesion flanges leads to softening of the adhesive so that the frictional connection is extensively destroyed. Under tension, displacement, loosening or slipping of the sealing layer from the adhesion flange can occur.

— Dirt, damp and low temperatures prevent a watertight connection of roofing panels to metal adhesion flanges.

Recommendations for the avoidance of defects

● The edge of the sealing layer should be formed from the roof membrane itself or by using a connecting foil made of a material of similar characteristics. A frictional connection of the pliable roof membrane with trims made of metal, asbestos cement or synthetic material should be avoided.

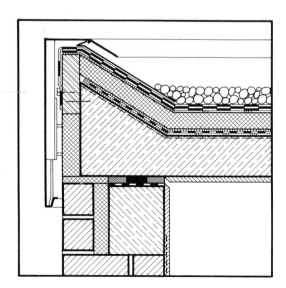

Solid slab flat roof
Edge of structural member

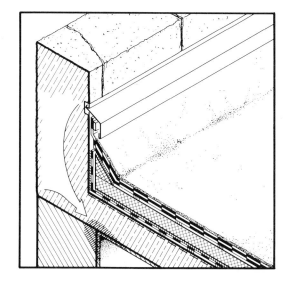

Defects to parapets frequently occur in the form of vertical cracks at intervals of a few metres. Simultaneously, dampness is discovered on the underside of the roof with this form of edge construction. Large metal copings on parapets become distorted and leak (see A 2.2.5).

If the roof membrane is fixed at a considerable distance above the angle fillet, overstressing occurs and cracks appear in the sealing layer at the outer corners (deflection points). Similarly, sealing layers which are either inadequately fixed or unstable slip and develop folds, ultimately working loose from their base.

Points for consideration

– A parapet is not necessarily the most appropriate termination, either functionally or structurally, for a solid slab roof that is not intended for traffic. Long-term solutions to this edge detail are difficult to construct.

– Parapets (and horizontal roof projections) are particularly exposed to climatic stress.

– Thermal stresses can be only marginally reduced by thermal insulation because of the large areas of cold surface. Thus the large thermal expansion must be provided for and the parapet constructed free of cracks by means of closely spaced expansion joints. Failing this, the width of cracks must be restricted by additional reinforcement.

– The large number of necessary expansion joints and the consequent increase in difficult sealing layer connections demand the most careful construction and supervision.

– Cladding units which are necessary for sealing but which are often large in area must be reduced to narrow panels with overlapping joints because of the solar gain.

– Sealing layers laid unprotected above parapets are subject to stress from ultra-violet rays as well as from thermal expansion of the parapet. The application of gravel or protective paint gives slight protection but only for a limited period.

– A parapet fixed directly to the roof slab forms a cold thermal bridge; this can be avoided only by internal insulation, which is not structurally desirable.

Recommendations for the avoidance of defects

● High parapets at the roof edge should preferably not be used.

● If a parapet is built it must have expansion joints of approximately 20 mm width at centres of ≤ 5 m. These expansion joints must be carefully sealed.

● The junction of the roof membrane with the parapet must be designed so that the anticipated movements of the parapet sections (especially in the area of the joints) do not lead to fracture or loosening of the roof membrane or at the connecting cover trim.

● Brickwork parapets should be reinforced.

● Parapets which are placed on the roof slab, held in position by their own weight, and anchored (e.g. precast reinforced concrete parapets), allowing consistent external thermal insulation of the roofing, are preferable to those fixed firmly to the roof slabs or external brickwork.

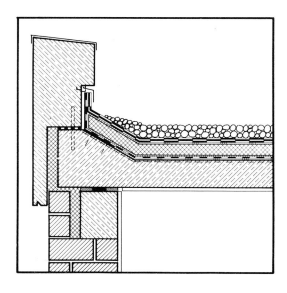

Problem: Bearing surface and expansion joints

In comparison with small unit roof coverings with continuous roof membranes, there are greater temperature stresses and different drying and bearing surface conditions in roofs of large area. The changes in form of the roof strongly stress the junction between roof and wall – the bearing surface. In loadbearing walls built of brick (the type mostly used in residential building) many cracks occur, resulting in dampness in the region of the bearing surface.

The construction of the bearing surface, and the design and execution of expansion joints in continuous roofing demand special care and attention during erection.

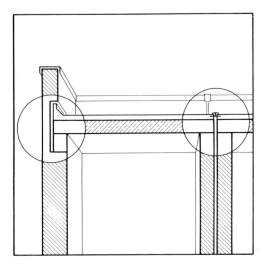

The problem of detailing the bearing of the roof slab on the structural wall and the expansion joints is complex, whether there is a rigid connection or a slip joint. Adjustments to accommodate varying spans and consequent different loading requirements, the depth of the slab, its fixing, continuity of reinforcement, selection of materials and construction techniques must all be taken into account. The decisions on bearing surfaces and expansion joints should therefore be made at an early stage of design in collaboration with the structural engineer.

On the following pages examples of defects in the design of bearings and expansion joints are examined.

Solid slab flat roof

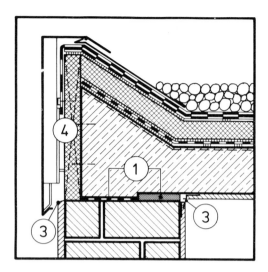

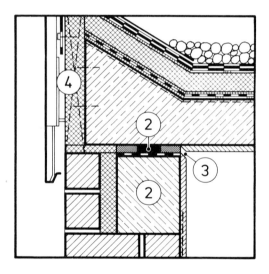

1 If the roof-bearing surface is designed as a rigid connection, a dividing foil should be laid in the bearing joint and a frictional connection on the inner third of the bearing wall surface avoided (see A 2.3.2).

2 If the expected linear expansion is to be accommodated in the slip joint between the wall and the roof, then a dowel mounting should be incorporated which will allow for this movement in the slip joint. Non-loadbearing walls should be separated from the roof slab by inserting a bituminous membrane between the two surfaces (see A 2.3.3).

3 Ceiling and wall plaster should be separated in the region of the slip joint. The outside face of the joint should be covered in a manner that prevents penetration by rainwater (see A 2.3.2 and 2.3.3).

4 The external vertical surface of the roof edge and the area surrounding the dowel should be protected by external quality thermal insulation of $\geqslant 1\cdot 3$ m² K/W. This should be continuous with the thermal insulation layer of the main roof (see A 2.3.3).

5 If expansion joints are provided to reduce the overall movement in the structural member, such a joint is to be continuous through the entire structure; otherwise, the roof slab should be designed to move at its support on a loose dowel in the region of the expansion joint (see A 2.3.4).

6 The expansion joint should be $\geqslant 20$ mm wide, should be protected by a pliable filler and carried through all the layers of the roof structure (see A 2.3.4).

7 In the sealing membrane, the expansion joint should be above any region of possible ponding and should be sealed by means of a preformed gasket. By this method, penetration of rainwater at the expansion joint should be prevented (see A 2.3.4).

8 The number of expansion joints within a roof surface should be limited to those essential (see A 2.3.4).

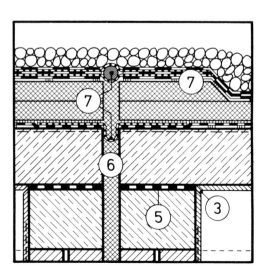

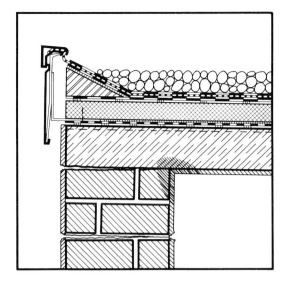

Roofs with short spans or minimal expansion and good external thermal insulation, in which no damage is to be expected and whose bearing brickwork needs no dowels or anchor straps, have still been found to show horizontal cracks at the bearing surface. These cracks are particularly clear where there is continuous wall plaster. If the area of the bearing surface externally is not protected by a fascia, rainwater penetrates. If the roof edge is not insulated, damp and discoloration appear at the corners of the room inside.

Points for consideration

– The sagging of reinforced concrete slabs produces distortion in the bearing surface which, with continuous bearing over the whole wall, can lead to eccentric loading and cause cracks in the bearing surface joint.

– A joint between the roof slabs and bearing surface brickwork is usually unavoidable because of sagging. Continuous plastering is not recommended if cracks are to be avoided.

– The roof upstand already forms a thermal cold bridge because of its vertical cold surface. The thermal insulation of the fascia and its connection with the horizontal thermal insulation of the roof structure are very important if low surface temperatures and the formation of condensation are to be avoided.

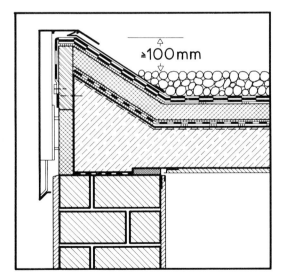

Recommendations for the avoidance of defects

● If roof-bearing surfaces can be made rigid, so that the expected linear expansion does not lead to damage, then a dividing foil should be laid in the bearing joint (e.g. by inserting a thermal insulating strip approx. 10 mm thick) so that a frictional connection in the inner third of the bearing surface is prevented.

● The ceiling and wall plaster should be separated in the area of the bearing joint. The outside face of the joint should be covered with a plaster stop to prevent penetration by rainwater.

● The external vertical surface of the roof edge should be protected with external quality thermal insulation with an insulating value of $\geqslant 1 \cdot 3$ m² K/W. This should be continuous with the thermal insulation layer of the roof structure.

Solid slab flat roof
Bearing surface and expansion joints

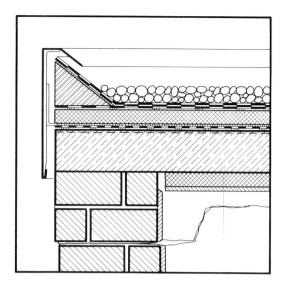

At the brickwork supports of solid roofs, horizontal and stepped cracks frequently occur; these lead to subsequent rain penetration. This damage may be found where there are no designed sliding or dowelled joints and the structural span is $\geqslant 12$ m (see A 1.1.3 − Linear expansion of roof slab).

This damage sometimes occurs despite dowels and sliding joints.

When fascia or dowels are uninsulated, damp and discoloration may occur in the room below.

Points for consideration

− If linear expansion of solid roofs is prevented by sliding joints, pressure builds up which, in members longer than 12 m, leads to fracture of the bearing brickwork. In such cases, sliding joints and dowels must be provided at the loadbearing wall junction, unless the linear expansion over the bearing surfaces is reduced by an increase in thermal protection or by expansion joints.

− The construction of dowels with a sliding joint between roof and bearing surface can lead to a tension-free dispersal of the expansion, if synthetic joints with low frictional resistance are used.

− The functioning of sliding foils is very dependent on the smoothness of the bearing surface. Unavoidable irregularities must therefore be corrected by increasing the thickness of the sliding foils and bearings. Expansion joints should be provided for structural members with large linear expansion, as internally in the living accommodation fully functioning slip joints are not easily incorporated in the domestic construction.

− Other roofing components which are rigidly constructed in close proximity to the slip joint are unable to accommodate the movements that occur; they therefore fracture above the joint, unless the total movement can be contained by the slip joint.

− Uninsulated roof edges and reinforced concrete dowels form thermal cold bridges which lead to the formation of condensation on the internal surfaces.

Recommendations for the avoidance of defects

● If the anticipated linear expansion is taken up by a sliding joint, a dowel should be located at the junction of the load-bearing wall and the roof slab, and sliding membrane inserted in the slip joint. Non-loadbearing walls should be separated from the roof slab by a bituminous membrane.

● The adjoining faces of the slip joints should have even surfaces. The sliding membrane used should be capable of countering the remaining friction by means of additional flexible strips.

● Wall and ceiling plaster should be separated at the junction of the slip joint. The wall joint should be protected so that rain cannot penetrate.

● The vertical edge of the roof and the dowel should have external thermal insulation of $\geqslant 1\cdot3$ m² K/W which must be continuous with the thermal insulation layer of the roof structure.

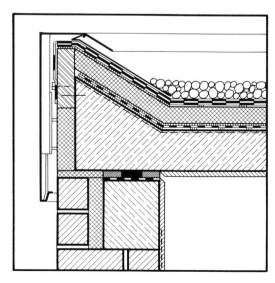

Solid slab flat roof
Bearing surface and expansion joints

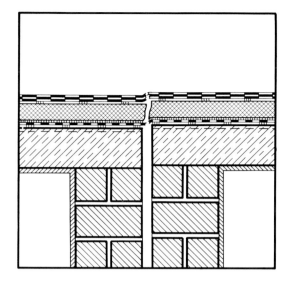

If solid roofs with structural members longer than 10 or 15 m (see A 1.1.3 – Linear expansion of roof slab) are not sub-divided by expansion joints and slip bearing, cracks are frequently found in the external brick walls.

In the expansion joints located in the roof slab, if the joint is not continuous through the loadbearing brickwork or the roof slab is not rigidly mounted, then the brickwork under the expansion joint fractures.

If expansion joints or the structural settlement joints are not continued through all layers of the roof structure as well as through the upstands, then these crack and cause dampness.

Points for consideration

– The overall extent of linear expansion of the roof slabs above the bearing can be effectively reduced by expansion joints.

– Expansion joints require carefully designed weathering detailing and may themselves cause defects in the flat roof. The presence of many expansion joints in the roof surface, especially expansion joints in areas of ponding, leads to increased risk of damage.

– In the area of the expansion joint the bearing brickwork as well as the roof structure is heavily stressed. If expansion joints in the roof slab and other movement joints in the structure are not considered structurally in the design of the bearing and the building of the roof structure, failure is inevitable.

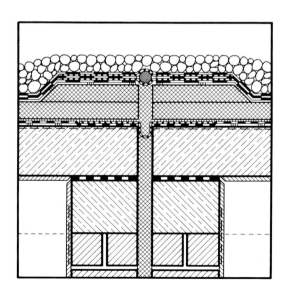

Recommendations for the avoidance of defects

● Expansion joints are arranged to reduce the interacting length of structural members. The joint should be continuous through the entire structure, or the roof slab should be slip jointed on a dowel at least in the region of the expansion joint.

● The expansion joint should be $\geqslant 20$ mm wide, should be protected from obstructions by a pliable filler and be continuous with the expansion provision through all the layers of the roof structure.

● In the sealing layer the expansion joint should be raised above the area of ponding and be formed by a raised gasket. Roof drainage over the expansion joints should be avoided.

● The number of expansion joints within a roof surface should be limited.

Problem: Drainage

On a flat roof drainage is of significantly greater importance than on a pitched roof which, with its easy dispersal of water and its gutters and downpipes at the eaves, has relatively few problems of water shedding. As the roof pitch is reduced the requirements of waterproofing and evenness of the roof membrane become greater, since the dispersal of the rainwater is slower and the sealing layer is exposed to standing water for a longer period of time.

Small unevennesses (caused by welts, for example) or slight sagging of the loadbearing structure can prevent the rainwater from flowing away easily.

In calculating the roof loading an additional factor must be considered: that of standing water which can result from clogged outlets or drainpipes. In the same way, additional stresses are transmitted to the sealing and thermal layers by ponding and by frequent temperature changes from wet and dry surfaces.

It is important at the building design stage to consider the discharge of rainwater from flat roofs and to locate and size outlets, etc. The selection of outlet patterns, gullies and downpipes is related to the structure of the roof for falls and to their connection with the sealing layer.

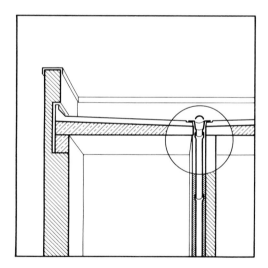

Only 8% of all the solid slab flat roofs investigated were drained by external gutters. No failures to the drainage elements were found – consequently they are not included in the following analysis of defects. The main areas of failure are found at internal gutters and individual outlets.

In parallel with the unfavourably located drainage elements on the roof surface, the connection to the sealing layers, as well as the installation of non-thermally insulated gullies and downpipes, are particularly liable to defects.

Solid slab flat roof

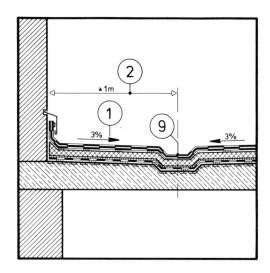

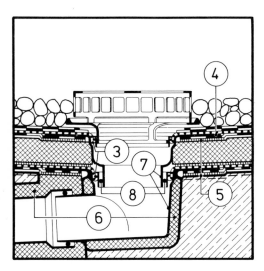

1 In principle, flat roof outlets should be located at the lowest part of the roof and drain the surrounding roof areas by means of an effective fall ($\geqslant 3\%$). Deflections, especially in light-weight structures and cantilevered slabs, must be taken into account (see A 2.4.2).

2 Individual outlets and gutters should be situated at a distance of $\geqslant 1$ m from roof abutments such as parapets, upstands, walls, etc. (see A 2.4.2).

3 Flat roof outlets must be constructed in a manner that ensures perfect connection to the roof membrane. They should be designed with a flexible flange which can be built into the multi-layer membrane surrounding the outlet (see A 2.4.3).

4 Special care should be taken to provide a good adhesion of adjacent sealing layers to the outlet flanges. The sealing layer should be secured only at the outer rim of the connecting flange, by means of adhesive (see A 2.4.3).

5 The increased thickness at the flanges of the outlet funnel, where the sealing layer is affixed, must be counteracted by a depression in the thermal insulation layer of the structural roof slab (see A 2.4.3).

6 Outlets and drainpipes should be lagged with thermal insulation material (see A 2.4.4).

7 In principle, the outlets should be composed of two sections if they are for single slab roofs. To accommodate the variable thickness of the insulating layer, they must be vertically adjustable (see A 2.4.4).

8 As protection against water seepage at the joint, a preformed groove and gasket seal should be provided between the outlet elements (see A 2.4.4).

9 Trough or channel gutters are preferable to box gutters (see A 2.4.4).

Solid slab flat roof
Drainage

Flat roof outlets may develop defects of function after a short time, especially if they are unfavourably located on the roof surface to be drained. In outlets which lie very close to the roof abutment this is most noticeable: the connections to the sealing membrane are not sealed, or the outlets become blocked by leaves, dirt or snow, so that the water can no longer flow away but lies in puddles on the roof surface.

The formation of puddles with subsequent cracking may also be discovered in roof surfaces of uneven depth, in which the outlets lie in the higher part of the roof.

Points for consideration

— Individual outlets and gutters can effectively drain rainwater only from that part of the roof surface which is laid to falls towards the gutters, etc. Standing water and consequent cracking in the area at the edge of the ponding must be avoided (see A 1.1.8 – Fall and the formation of ponding on the roof membrane).

— In the design of the drainage elements possible deflection should be taken into account in lightweight structures and cantilevered slabs, to ensure that the outlets lie in the deepest areas of the roof.

— Outlets and gutters must be fastened to the roof membrane by means of connecting elements (flanges) and therefore need adequate space for their construction.

— If outlets and gutters are located close to roof abutments, parapets or upstands, leakage at the junction with the roof membrane is likely, as effective waterproofing is made difficult, if not impossible, by lack of working space. Additionally, there is the danger of ice forming at the outlet, so that installation of heated drainage elements may become necessary at very exposed sites.

— Wind causes leaves, dirt and snow to collect, especially in the corners and channels of roof surfaces, and these prevent drainage by obstructing the water flow to outlets and gutters.

Recommendations for the avoidance of defects

● In principle, flat roof outlets must lie in the lowest areas of the roof surface, with an effective fall ($\geqslant$ 3%) of the drainage area.

● Individual outlets and gutters should be at a distance of $\geqslant$ 1 m from roof upstands such as parapets, walls, etc.

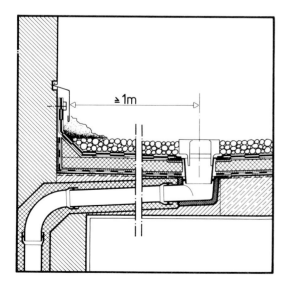

$\geqslant$ 1m

Solid slab flat roof
Drainage

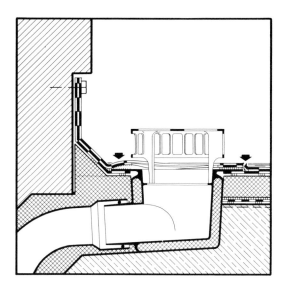

In drainage away from the edge of the roof, drainage channels, as well as individual outlets (gullies), have been found to show defects of construction at the connection to the sealing layer. Incorrect adhesion between the sealing membrane and preformed metal sections causes leakage through the roof structure and damages the ceiling below by partially loosening the plaster.

Points for consideration

– Because of their function and position in the lowest part of the roof surface, drainage channels are subjected to strong stresses from rainwater. Leakages therefore have a particularly damaging effect.

– Effective flat roof outlets close to, or even built into, parapets or roof edge upstands are very difficult to construct (see A 2.4.2 – Arrangement and position of drainage elements).

– Where there is a large variation of temperature between different materials (e.g. bituminous roof membrane – cast iron roof outlet) considerable stresses arise which must be balanced in the transition areas (e.g. by the use of flexible connecting layers).

– Overlaps and strengthening layers in the area of the outlet flange produce an increase in depth at the outlet rim, impeding the flow of rainwater. This results in the formation of puddles.

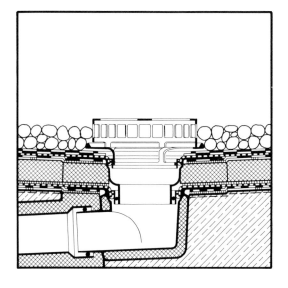

Recommendations for the avoidance of defects

● The flat roof outlet must be so constructed as to ensure a satisfactory connection to the roof membrane.

● Rigid connections without stress compensation should not be made. Gullies in flat roofs should be provided with an adequate flexible connecting flange in every drainage plane.

● Special care should be taken to provide good adhesion of the sealing layer, and also the vapour barrier, to the appropriate adhesion flange. To allow for movement, the sealing layer should be secured only in the outer area of the connecting flange.

● Overlapping in the area of connection of the roof membrane with the drainage outlet results in increased thickness. This must be compensated by a depression in the thermal insulation layer or concrete roof.

Solid slab flat roof
Drainage

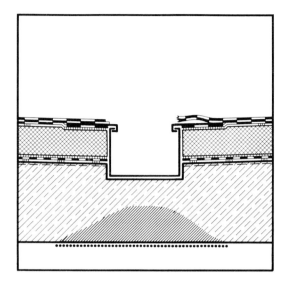

Damage to the drainage elements caused by dampness occurs particularly where drainpipes, gullies or channels have been installed without thermal insulation lagging, and where incorrect outlets or defectively constructed channels have been built into the structure. After the defective sealing connections already mentioned, these errors have been found to be the most frequent cause of failure in flat roof drainage.

Points for consideration

- Roof outlets and the connected drainpipes present problems of thermal technology as well as of noise. Where uninsulated outlets and pipes are installed, condensation forms on the under-side of the roof and on the drainpipes. This results in dampness to the structure (thermal insulation, concrete roof and walls).

- Roof surfaces must be effectively drained during the course of construction work to minimise damp penetration into the structure. The vapour barrier layer should be connected to the outlet.

- As the structural depth of a roof structure can vary, roof outlets must be adjustable.

- Where drainpipes are blocked or outlets are too narrow in cross-section, water must not be able to penetrate the roof structure through backwash.

- Drainage channels of preformed sheet metal present an exceptional hazard because of the arrangement of sliding seams and folds at their junctions.

Recommendations for the avoidance of defects

● Outlets and drainpipes should have thermal insulation lagging. The installation of heated roof gullies in frost-prone areas provides an additional safeguard against frost damage to a flat roof, as these parts can be kept free of ice (not normally installed in the British Isles, but common on the continent of Europe).

● In principle, outlets composed of two sections should be provided for flat roofs: these can be adjusted vertically to the varying thicknesses of the insulating layer.

● For protection against water seepage at the joint, a preformed groove and gasket seal should be provided between the outlet elements.

● Box gutters should not be used, because of their difficult and insecure connections. Trough or channel gutters which can be formed directly from the roof membrane are recommended.

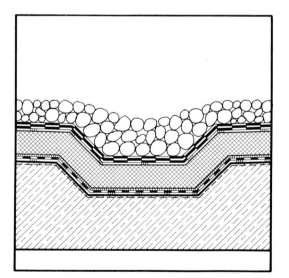

Solid slab flat roof
Points of detail

General texts and principles

Eichler, Friedrich: Bauphysikalische Entwurfslehre, Band 2, 4. Auflage, Verlagsgesellschaft Rudolf Müller, Köln 1973.

Hoch, Eberhard: Flachdächer – Flachdachschäden, Verlagsgesellschaft Rudolf Müller, Köln 1973.

Jungnickel, Heinz u.a.: Abdichtungs- und Bedachungstechnik mit Kunststoffbahnen, Verlagsgesellschaft Rudolf Müller, Köln 1969.

Meyer-Bohe, Walter: Dächer, Verlagsanstalt Alexander Koch GmbH, Stuttgart 1972.

Moritz, Karl: Flachdachhandbuch – flache und flachgeneigte Dächer, 4. Auflage, Bauverlag Wiesbaden und Berlin 1975.

Schild, E.; Oswald, R.; Rogier, D.: Bauschäden im Wohnungsbau Teil II – Bauschäden an Dächern, Dachterrassen, Balkonen – Ergebnisse einer Umfrage unter Bausachverständigen, Verlag für Wirtschaft und Verwaltung – Hubert Wingen, Essen 1975.

Zentralverband des Dachdeckerhandwerks: Richtlinien für die Ausführung von Flachdächern, Ausgabe Januar 1973, Helmut Gros Fachbuchverlag, Berlin 1973.

DIN 1045 – Beton- und Stahlbetonbau, Januar 1972.

DIN 1053 – Mauerwerke, Berechnung und Ausführung, Blatt 1, November 1974.

DIN 1055 – Ergänzende Bestimmung zu Blatt 4, Lastannahmen im Hochbau, Verkehrslasten, Windlasten; Ausgabe Juni 1939, März 1969.

DIN 4122 – Abdichtung von Bauwerken gegen nicht drückendes Oberflächenwasser und Sickerwasser mit bituminösen Stoffen, Metallbändern und Kunststoffolien, Richtlinien, Ausgabe Juli 1968.

DIN 18530 – Massive Deckenkonstruktionen für Dächer, Dezember 1974.

Connection to vertical abutments

Buch, Werner: Wärmedämmung von Dächern mit Hartschaum aus Styropor. Herausgeber: Informationszentrum Styropor.

Haushofer, Bert: Ein Flachdach ist stets so gut wie seine Anschlüsse. In: Deutsches Dachdeckerhandwerk (DDH), Heft 10/68, S. 571–575.

Hummel, J.: Abschlüsse und Anschlüsse im flachen Dach. In: Deutsches Dachdeckerhandwerk (DDH), Heft 14/69, S. 870–871.

Kakrow, K. H.: Detailbildung in der Flachdachabdichtung. In: Deutsches Dachdeckerhandwerk (DDH), Heft 1/68, S. 25–30.

Schaupp, Wilhelm: Das bituminöse Dach ist stets so gut oder so schlecht wie seine Anschlüsse. In: Deutsches Dachdeckerhandwerk (DDH), Heft 19/69, S. 1267–1282.

Schlenker, Herbert: Die Fachkunde der Bauklempnerei, A. W. Geutner Verlag, Stuttgart 1971.

Edge of structural member

Haushofer, Bert: Ein Flachdach ist stets so gut wie seine Anschlüsse. In: Deutsches Dachdeckerhandwerk (DDH), Heft 10/68, S. 571–575.

Hummel, J.: Abschlüsse und Anschlüsse im flachen Dach. In: Deutsches Dachdeckerhandwerk (DDH), Heft 14/69, S. 870–871.

Kakrow, K. H.: Detailausbildung in der Flachdachabdichtung. In: Deutsches Dachdeckerhandwerk (DDH), Heft 1/68, S. 25–30.

Lorenzen, Heinz: Auswertung der Erfahrungen aus der Sturmkatastrophe. In: Deutsches Dachdeckerhandwerk (DDH), Heft 4/69, S. 194–197.

Österreichisches Institut für Bauforschung: Abdichtungen und Abläufe bei Flachdächern, Dachterrassen, Balkonen, Loggien, Naßräumen. Forschungsbericht 39, 2. Auflage, Wien 1973.

Pfefferkorn, Werner: Konstruktive Planungsgrundsätze für Dachdecken und ihre Unterkonstruktionen. In: Das Baugewerbe, Heft 18/73, S.57–65, Heft 19/73, S. 54–59, Heft 20/73, S. 86–90, Heft 21/73, S. 54–63.

Praktisch, Theo: Flachdachanschlüsse mit Bitumendachbahnen. In: Deutsche Bauzeitschrift (DBZ), Heft 4/71, S. 689–690.

Rick, Anton W.: Sturmschäden in den USA. In: Das Dachdeckerhandwerk (DDH), Heft 9/70, S. 557–562.

Schaupp, Wilhelm: Das bituminöse Dach ist stets so gut oder so schlecht wie seine Anschlüsse. In: Deutsches Dachdeckerhandwerk (DDH), Heft 19/69, S. 1267–1282.

Schlenker, Herbert: Die Fachkunde der Bauklempnerei, A. W. Geutner Verlag, Stuttgart 1971.

Timmerberg, Carl H. u.a.: Details ... Details..., Praktische Hinweise zu Esser-Produkten, Düsseldorf 1974.

Bearing surface and expansion joints

Balkowski, F. D.: Die Rißbildung am Deckenauflager. In: Das Dachdeckerhandwerk (DDH), Heft 2/74, S. 88–91.

Brandes, K.: Dächer mit massiven Deckenkonstruktionen – Ursache für das Auftreten von Schäden und deren Verhinderung. In: Berichte aus der Bauforschung, Heft 87, Verlag Wilhelm Ernst & Sohn, Berlin 1973.

Buch, Werner: Das Flachdach, Dissertation, Darmstadt 1961.

Grunau, Edvard B.: Probleme des Flachdachs. In: Deutsche Bauzeitung (db), Heft 11/73, S. 1262–1266.

Kakrow, Helmut: Das flache Dach – Planung und Ausführung am Beispiel der Praxis. In: Deutsches Dachdeckerhandwerk (DDH), Heft 4/66, S. 144–147, Heft 9/66, S. 436–440.

Kramer-Doblander, Herbert: Temperaturspannungen in Flachdachkonstruktionen. In: Bitumen, Teere, Asphalte, Peche..., Heft 3/70, 21. Jahrgang, S. 93–96.

Moritz, Karl: Technische und bauphysikalische Notwendigkeiten bei Flachdächern. In: Deutsches Architektenblatt (DAB), Heft 9/70, S. 315–318.

Pfefferkorn, Werner: Konstruktive Planungsgrundsätze für Dachdecken und ihre Unterkonstruktionen. In: Das Baugewerbe, Heft 18/73, S. 57–65, Heft 19/73, S. 54–59, Heft 20/73, S. 86–90, Heft 21/73, S. 54–63.

Planckh, Rudolf: Die erforderliche Gleitfähigkeit der Flachdachschichten. In: Bitumen, Teere, Asphalte, Peche..., Heft 10/72, S. 408–410.

Probst, Raimund: Dachdecken. In: Das Bauzentrum, Heft 4/68, S. 13–17.

Rick, Anton W.: Fugenteilungen in Flachdächern. In: Deutsches Dachdeckerhandwerk (DDH), Heft 20/69, S. 1338–1340.

Zimmermann, Günter: Wärmedämmung bei nicht belüfteten Flachdächern. In: Deutsches Dachdeckerhandwerk (DDH), Heft 10/68, S. 597–602.

Drainage

Buch, Werner: Das Flachdach, Dissertation, Darmstadt 1961.

Haefner, R.: Die Abdichtung von unterkellerten Hofdecken, Terrassen über Nutzräumen und Flachdächern. In: Boden, Wand und Decke, Heft 7/66, S. 588–696.

Österreichisches Institut für Bauforschung: Abdichtungen und Abläufe bei Flachdächern, Dachterrassen, Balkonen, Loggien, Naßräumen; 2. Auflage, Wien 1973.

Soyeaux, H.: Rinnenabdichtung mit Polyisobutylen-Dachbahnen. In: Deutsches Dachdeckerhandwerk (DDH), Heft 18/69, S. 1158–1164.

Voorgang, H. J.: Die Entwässerung flacher Dächer I–IV. In: Deutsches Dachdeckerhandwerk (DDH), Heft 1/69, S. 14–16; Heft 3/69, S. 130–132; Heft 6/69, S. 278–282; Heft 9/69, S. 532–535.

Problem: Sequence of layers and individual layers

The structural double slab roof is characterised by a ventilated space between the sealing layer which provides weather protection (upper slab) and the thermal insulation layer which serves to retain warmth (lower slab). The double slab roof has, in comparison with the single slab roof, fewer structural layers; however, for its correct functioning a constant exchange of air must be maintained between the enclosed roof space and the air outside. The change of air is dependent on the cross-ventilation, passage of air, location of the building and the thermal conditions. Double slab roofs are thus not practicable for all types of building; their construction is complex and consequently they are prone to further problems.

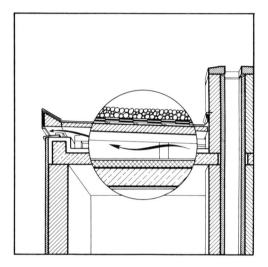

Double slab ventilated roofs for residential buildings are mainly of two construction types:

— the double slab roof with solid reinforced concrete roof for the loadbearing slab and a weathering flat or inclined light-weight skin on timber framing;

— the light flat double slab roof which consists of a timber structural frame with a timber boarded weathering surface.

The following recommendations refer to these two design forms.

Parallel with the failure to the sealing layer which occurs in all flat roofs, the primary cause of failure in the cross-section of the double slab roof is condensation which can be traced back to defective ventilation. On the following pages this defect, as well as other failures, is illustrated and analysed and the conclusions are tabulated as recommended practice.

Double slab flat roof

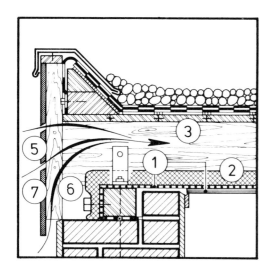

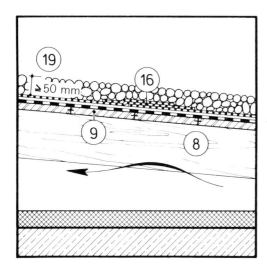

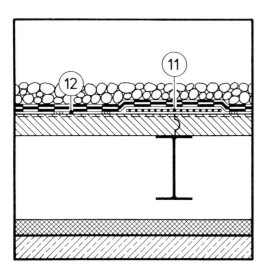

1 The lower slab of the double slab roof should have a vapour permeance value (diffusion resistance equivalent air layer thickness) of $\geqslant 10$ m and should be designed in such a way that there is no direct connection between the roof space and the room space below. With lightweight structural forms (e.g. timber lattice beams or joists), foils with overlap joints and carefully designed edge connections should be arranged under the thermal insulation (see B 1.1.2).

2 A thermal insulation layer with a heat resistance of $\geqslant 1 \cdot 3$ m² K/W should be applied continuously on the under-side of the lower skin. Loose-laid light thermal insulation fills, which could be blown away by a gust of wind, should not be used (see B 1.1.3 and 1.1.4).

3 The upper slab should have a high heat-storage capacity – timber boarding with multilayer bitumen coverings is preferable to sheet-metal profiles; however, materials which are unstable in damp conditions should not be used (see B 1.1.3).

4 The air space should be $\geqslant 100$ mm at its lowest point – considerably greater depth (300–600 mm) and a fall of $\geqslant 5°$ (9%) should be aimed at in the upper surface (see B 1.1.5).

5 Double slab ventilated roofs should have openings for ventilation providing a free air cross-section of $\geqslant 0 \cdot 2\%$ (1/500) of the roofing surface on at least two opposite sides of the roof (see B 1.1.6).

6 The openings must be arranged so as to ensure a uniform ventilation of the whole roof surface. Continuous vents are preferable to individual openings. The openings and the depth of air space must be designed and constructed in such a way that blocking by insulating material or plaster is avoided (see B 1.1.4 and 1.1.6).

7 The junction between the opening and the air space should be as direct as possible, but allowing for protection from rainwater (see B 1.1.6).

8 The structural members of the double slab should be designed so that bending, even under additional loading, does not lead to unacceptable stressing of the roof or to minimising of the fall (see B 1.1.7).

9 On the double slab flat roof only materials which are stable under damp conditions should be used. This applies to slabs made of wood or straw fibre material for the roof decking (see B 1.1.7).

10 The built-up roof membranes should be fixed to the decking by spot or strip adhesion, or nailed if the base is suitable (see B 1.1.8).

11 With large base slabs, such as aerated cement roof slabs or precast hollow cement boards, in addition to spot adhesion the roofing sheets should remain free for a width of approximately 300 mm over the expansion joints of the slabs (transverse joints). This can be achieved by laying a loose strip over the expansion joint (see B 1.1.7 and 1.1.8).

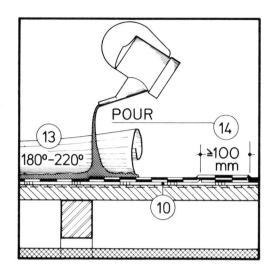

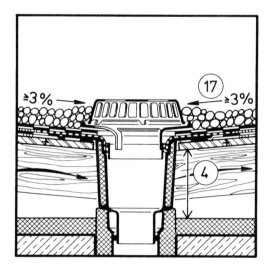

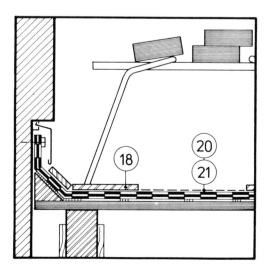

12 The composition of the base must permit a satisfactory application of the sealing layer. It must be smooth, dry, not too porous and be free of oil or paint residue (see B 1.1.8).

13 Heat sealing – according to the type of bitumen at 180°–230°C – should be carried out only in dry weather and at outside temperatures of ≥ 5°C (see B 1.1.9).

14 With sealing layers made of bituminous roofing felt the longitudinal and transverse covering seams should be ≥ 100 mm wide. The pouring and rolling process is basically preferable to other methods of adhesion, since it offers the greatest guarantee of vacuum-free and even adhesion (see B 1.1.9).

15 Roofing sheets with metal linings may only be used in areas of small variation in temperature, e.g. under gravel fills and terrace coverings (see B 1.1.9).

16 Single-layer loose-laid synthetic roofing sheets should be given the additional protection of cover strips as well as adhesion or welding of the seams (see B 1.1.9).

17 To avoid the formation of puddles, especially where the surface protective layer may be defective, the fall of the upper slab should be approx. 1·4° (3%) to the drainage outlets, if a fall of 5° (approx. 9%) is not already provided for ventilation (see B 1.1.10).

18 Repair work to walls abutting the roof structure, in which the sealing layer is often unprotected, may be carried out only after appropriate precautions have been taken, e.g. board and plank bases preventing pressure points from scaffolding supports and damage through movement of gravel (see B 1.1.11).

19 Basically, every flat roof should have an upper surface protective layer. A gravel fill, which is practicable for a roof fall of ≤ 5°, should be to a depth of 50 mm, of washed round granules with a coarseness of 16–32 mm dia, and should be laid loosely and uniformly. The loading resulting from this upper surface protective layer must be taken into account in the calculations for the loadbearing structure with regard to depth, span width and bending (see B 1.1.12).

20 Roof surfaces which, because of steep falls or lower loadbearing capacity of the upper shell, do not permit loose gravel fill as the protective layer, should be protected by chippings, sandings or painting, according to the fall, loadbearing capacity and the material of the sealing layer (see B 1.1.12).

21 Thin upper surface protective layers (reflective coatings, sandings, etc.) must be inspected for maintenance and if necessary renewed at frequent intervals (see B 1.1.12).

Double slab flat roof
Sequence of layers and individual layers

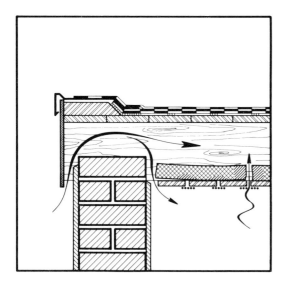

With double slab roofs made up of lightweight structured members, the following types of damage have been observed, caused by gaps in the surface of the supporting frame or at the wall connections:

Wall connections of the lower slab which are not watertight – especially along the external walls in the area of the ventilation openings – lead to internal stressing. In the area of the gaps the wall surface shows partial darkening as a result of dust deposits and thermal variance.

If, through gaps in the lower slab, the air of the room space is directly connected with the roof air space, or if the ventilation of the outlet pipes ends in the intermediate roof space, this results in heavy condensation and dampness to the thermal insulation and to the timber structure.

Points for consideration

— Where there is a direct connection between the intermediate roof space and the room space, so that the warm, damper air inside penetrates into the roof space (because of excess pressure in the inner space), or where the lower slab offers no resistance to the diffused water vapour in the room below, the ensuing condensation can no longer be dried out by the change of air in the roof space – particularly in the case of damp rooms or rooms with mechanical ventilating or air-conditioning plant.

— Certain wind conditions result in a pressure differential between roof space and room space, which, if there are gaps, can lead to stress through the lower lightweight slab. With the ventilated double slab roof, the lower slab must also fulfil a wind sealing function.

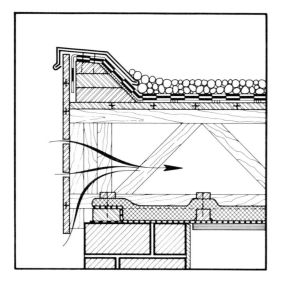

Recommendations from the avoidance of defects

● The lower slab of the double slab roof should have a vapour permeance value (diffusion resistance equivalent air layer thickness) of $\geqslant 10$ m and should be designed in such a way that no exchange of air is possible between roof space and room space.

● In examples of lightweight lower slabs (e.g. timber joists), foils with overlap joints and carefully designed edge connections must be formed under the thermal insulation.

Double slab flat roof
Sequence of layers and individual layers

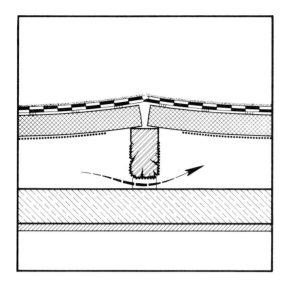

In a number of double slab flat roofs with poor or defective ventilation the upper slab, made of light timber or straw fibre slabs, has been found to be the only thermal insulation in the cross-section of the roof. The result is condensation in the roof space, swelling, rotting and sagging, leading to extensive loss of loadbearing capacity.

Points for consideration

— With adequate ventilation, a thermally insulated upper slab of the double slab roof offers little protection against loss of heat in the room space. Low temperatures in the lower slab, and consequently condensation, have to be contended with.

— With defective ventilation, a double slab roof with thermally insulating upper slab acts like a single slab roof with a defective vapour barrier. Condensation then occurs in the upper slab and in structural parts.

— With rapid changes in temperature (thunder showers, night cooling) condensation can occur, especially on upper slabs with low heat-storage capacity. Materials which are not stable in water can be damaged under this stress.

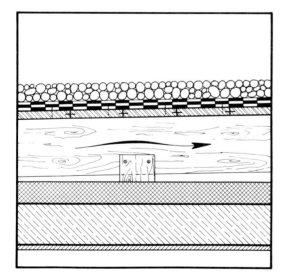

Recommendations for the avoidance of defects

● A thermal insulation layer having heat resistance of $\geqslant 1\cdot 3$ m² K/W should be applied continuously to the lower slab.

● The upper slab should have a high heat-storage capacity — timber boarding with multilayer bituminous coverings is preferable to metal-sheet cover; however, materials which are unstable in damp conditions should not be used.

Double slab flat roof
Sequence of layers and individual layers

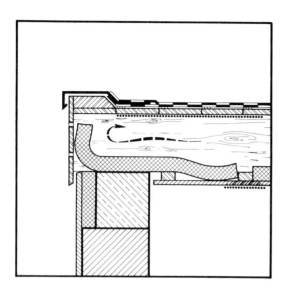

Where the thermal insulation of double slab ventilated roofs is not continuous over the lower slab, the defective areas are characterised by damp dark stains, particularly in roofs with lightweight plaster ceilings.

Where the thermal insulation is simply inadequate, this damage occurs over a large area.

If the ventilated openings or the air space between slabs is blocked by the thermal insulation, damp penetration leads to decay in the upper and lower slabs and in the loadbearing timber beams.

Points for consideration

— Gaps in the thermal insulation, due to negligent construction or difficult laying conditions, or to the lightweight insulating material being shifted by the wind, form thermal bridges which lead to surface condensation on the underside of the roof.

— Small or narrow individual ventilation openings and air spaces located close to the wall and the fascia board can easily be blocked during laying of the thermal insulation. Effective ventilation to the roof space is thus adversely affected.

Recommendations for the avoidance of defects

● The thermal insulation provided to the lower slab of the ventilated double slab roof should exceed the minimum specified heat resistance requirement of $1·3$ m² K/W.

● The thermal insulation must be applied continuously to the lower slab. The work of laying should be carefully carried out and supervised. An insulating material appropriate to the constructional situation should be selected so that uniform application is possible without difficulty. Loose, lightweight thermal insulation fills which can be shifted by wind should not be used.

● The ventilation openings and the depth of the air space must be designed and constructed in such a way that blocking by insulation material is avoided.

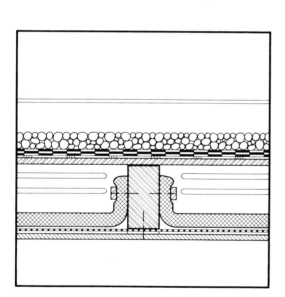

Double slab flat roof
Sequence of layers and individual layers

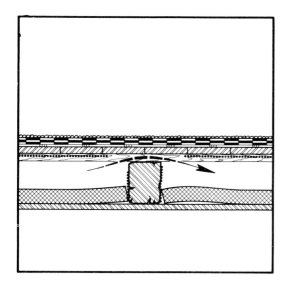

If the air space in a double slab roof is only a few centimetres deep or is interrupted or reduced by channel gutters and roof structure, or if the direction of stress of the roof joist (e.g. in terraced houses) or the form of the loadbearing structure does not allow for free ventilation between openings on opposite sides of the structural member, the result is condensation in the roof space. Rotting and fungus in the timber boards, joists, etc., result in a corresponding loss of loadbearing capacity in the structure.

Points for consideration

— With the double slab roof a constant exchange of air between the whole roof space and the outside air is necessary.

— Changes of air can be produced by wind forces and, in the case of sloping roofs or deep air space, by thermal upthrust. As effective wind ventilation is subject to variation caused by climate and position, double slab roofs which are constructed with steep falls or have a deep air space are preferable for ventilation.

— Upthrust and wind can only be effective for the ventilation of the whole surface of the roof space when the pressure differentials allow for the entry and discharge of air by adequate dimensioned openings on opposite sides of the roof structural space. There should be no high aerodynamic resistances to the air flow in the cross-section of the roof. Long air passages which change direction, small openings and narrow cracks present resistance to air flow.

— In contrast to the single slab roof, the ventilated double slab roof cannot be used for every type of roof structure and every design and building situation.

Recommendations for the avoidance of defects

● Double slab ventilated roofs must have, on at least two opposite sides, ventilation openings with unobstructed cross-section of 0·2% (1/500) each of the roof surface.

● The air space should have a depth of ≥ 100 mm at the lowest point — considerably greater depths (300–600 mm) and a fall of the upper shell of ≥ 5° (approx. 9%) should be achieved.

● The total air space of the roof must be ventilated; partial zones without cross-ventilation must be avoided.

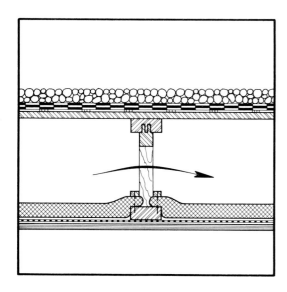

Double slab flat roof
Sequence of layers and individual layers

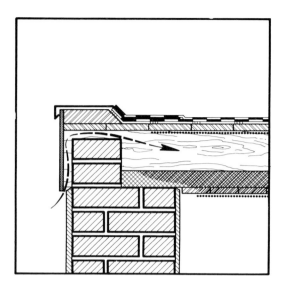

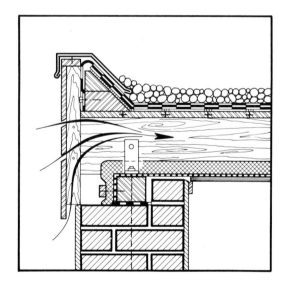

Condensation dampness in the double slab roof occurs mainly through inaccurate dimensioning and location of the ventilation openings.

The openings may be insufficient, consisting of a few checker bricks or borings which ventilate only part of the enclosed joist area, and connected with the air space by narrow angled channels and small pipes which are formed by a proprietary 'ventilator' system built into the upper slab, or are blocked by fill material or plastered over. The presence of dampness particularly in the timber parts of the roof – joists, frames and boarding – results in serious damage to the whole structure. Damage caused by over-dimensioned ventilation openings has not been found.

Points for consideration

– The ventilated double slab roof only functions under constant change of air between the roof space and the external air.

– A constant change of air presupposes large, uniformly distributed ventilation openings on at least two opposite sides of the building and a simple direct air circulation.

Recommendations for the avoidance of defects

● Double slab ventilated roofs must have – on at least two opposite sides – ventilation openings with an unobstructed cross-section of $\geq 0.2\%$ (1/500) of the roof surface.

● The ventilation openings must be located to provide a uniform ventilation of the whole roof surface. Continuous slits are preferable to individual openings. They should be so positioned as to preclude blocking by fill material or plaster.

● The connection between opening and air space – allowing for protection from rainwater – should be as direct as possible.

Double slab flat roof
Sequence of layers and individual layers

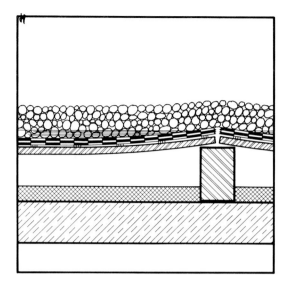

In a large number of double slab roofs with a lightweight upper slab, changes in form – primarily sagging of the timber boards and deformation of the loadbearing joist – lead to the formation of troughs and cracks in the roof membrane. The deformations are due either to overloading (e.g. deep gravel fill) or to loss of strength as a result of dampness.

Where the roof has an upper slab made of large slabs (precast cement boards), firmly adhering sealing layers or those that have been laid without sufficient tolerance crack above the transverse and longitudinal joints of the slabs.

Points for consideration

– Even a gravel fill can lead to overloading, especially on lightweight boarding.

– On the upper slab of a double slab roof provided with good ventilation, sudden cooling can lead to the formation of condensation. Through this condensation, and also through rainwater penetration which gets into the boarding during storage or through small gaps, the whole roof can be completely destroyed.

– The movement and sagging of large roof slabs can overstress the roof membranes at the transverse joints of the slabs when these are evenly bonded.

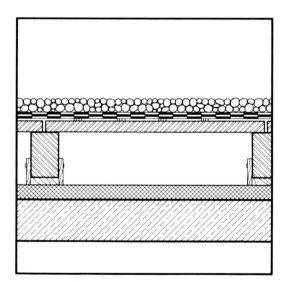

Recommendations for the avoidance of defects

● The structural parts of the double slab flat roof should be designed in such a way that sagging, even under additional loading, does not lead to unacceptable tensioning of the roof or to reversal of the fall.

● On the double slab flat roof only material which is stable in damp conditions may be used. This applies particularly to boarding material made of timber or straw fibre.

● The roof membrane should remain unbonded for a width of approximately 300 mm, particularly over the transverse joints of large roof slabs.

Double slab flat roof
Sequence of layers and individual layers

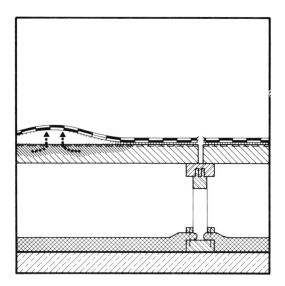

Bituminous sealing strips and synthetic foils which have been stretched and fixed tightly to the base frequently show cracks, especially in the joint area of large slabs, where the sealing base consists partly of aerated concrete slabs, precast concrete planks and chipboard.

Points for consideration

– Sealing panels bonded evenly to the base are committed to movements of the base and can thus be overstressed. On large-area bases such as aerated concrete roof slabs or precast concrete planks, these movements are considerable, especially where no temperature-reducing surface protective layer is provided.

– Spot or strip fixing of the lower layer reduces these stresses as the expansion of a great proportion of the area is absorbed.

– The reduced weathering quality of this form of construction can, in roofs with a fall of $\leqslant 5°$, be neutralised by applying a gravel fill ($\geqslant 50$ mm deep and of 16–32 mm dia) (see also B 1.1.12 – Climatic demands on the sealing layer – Upper surface protection). If this is not possible for structural reasons, then the roof membrane must be laid so that it can withstand bad weather.

– The trapping of dampness and air under the roof membrane, caused by a damp and dirty base, develops under strong sunshine to overpressure and, where there is no compensating capacity, to the formation of bubbles.

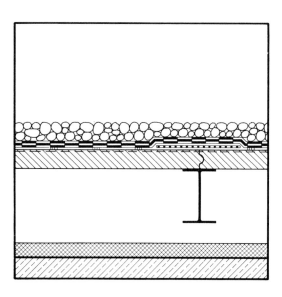

Recommendations for the avoidance of defects

● The roof membrane should be bonded by spot or strip adhesion to the base, or nailed if this is possible.

● With large-area base slabs, such as aerated concrete slabs, in addition to spot adhesion the roof membrane should remain unbonded over the expansion joints of the slab (transverse joints) for a width of 300 mm. This can be achieved by inserting a loose strip.

● The composition of the lower structure must permit uniform laying of the sealing panels. It must be smooth and dry, free from oil or paint residue, but not too porous.

Double slab flat roof
Sequence of layers and individual layers

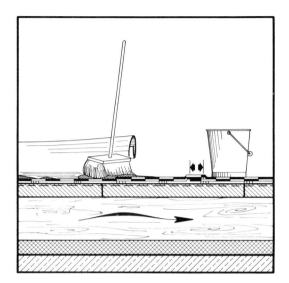

Sealing damage with consequent dampness or rotting of materials which are not stable under damp conditions, e.g. chipboard or straw fibre slabs, is often caused by leakages resulting from defective construction of the sealing layer. The defects occur mainly with built-up bituminous membranes, but they have been known to occur with synthetic foils. On occasions the overlap joints loosen because of insufficient width of overlap or defective bonding, or there is inadequate adhesion of the layers to each other or to the base (see B 1.1.8 – Laying of roof membrane on the base).

Points for consideration

A flat roof, particularly one with no fall, made of materials which are not stable in damp conditions requires very careful construction of the sealing layer.

– The weatherproofness of strip coverings depends on the careful construction of the overlap joints and, with built-up sealing layers, on the even adhesion of the layers.

– The trapping of impurities, water and air prevents an even bonding of the layers and can lead to water penetration on surfaces which are not protected from the sun's rays and are affected by changes in the external climate.

– Unprotected sealing layers with metal foil linings undergo large thermal expansions which can lead to the loosening of adhesive bonding at the overlap joints.

– Wet weather and external temperatures $< +5°C$ make bonding of the sealing layer more difficult.

Recommendations for the avoidance of defects

Sealing work on the flat roof should be carried out with special care and under strict control:

● With sealing layers made of bituminous roofing felt the longitudinal and transverse overlap seams should be $\geqslant 100$ mm wide. The layers should be evenly bonded to each other. The pouring and rolling process is basically preferable to other adhesion techniques, since it offers the greatest possible guarantee of vacuum-free and even adhesion.

● Heat sealing – at 180°–220°C according to the type of bitumen – should be carried out only in dry weather and at external temperatures of $\geqslant +5°C$.

● Roofing panels with metal linings may be used only in areas of low variation in temperature, e.g. under gravel fill and terrace finishes.

● Single-layer loose-laid synthetic roofing panels should be given added protection by means of cover strips, in addition to bonding or welding of the overlap seams.

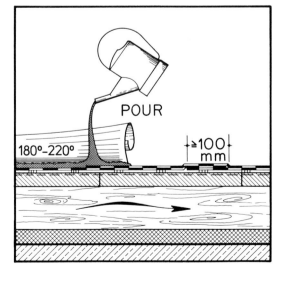

Double slab flat roof
Sequence of layers and individual layers

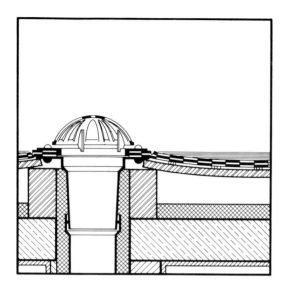

Puddles formed on the surface of roofs with no fall, as a result of deformation of the upper lightweight boards or the timber roof joists, lead to the formation of cracks in the roof membrane at the edge of the puddle and resulting dampness. The same effect has been observed where roof outlets are too high or channel gutters have been installed with no fall.

Particularly in roofs with no fall, leakages in the roof surface or at the connections cause serious dampness.

Points for consideration

— Roofs with no fall are always liable to constant standing water as the rain cannot flow away rapidly. A large quantity of water can penetrate through even a small leak.

— A roof cannot be constructed completely flat. Either the loadbearing slab itself reveals depressions caused in manufacture or by deformation, or the roof membrane is not level because of overlaps. In a roof with no fall, puddles are unavoidable.

— Ponding can be expected if lightweight slabs are excessively deflected as a result of increased loading (e.g. gravel fill) and where the drainage outlets are not located in the depressions of the roof surface.

— The roof membrane is thicker in the area of the adhesion flange of the roof outlets. If this greater depth is not allowed for, the edge of the outlet lies above the level of the roof; this results in the formation of puddles.

— The wet and dry areas of an unprotected roof membrane are simultaneously exposed to variable stresses. At the edge of the ponding stresses arise which, together with drying dirt deposits, lead to cracks. In the area of constant standing water these cracks have serious consequences.

Recommendations for the avoidance of defects

● If the upper slab, and consequently the sealing layer, does not (for technical reasons of ventilation – see B 1.1.5 – Ventilation) have a fall of 5° (9%), to avoid ponding the fall to the outlets should be ⩾ 1·4° (3%), particularly if no surface protective layer is provided to control the damp.

● The deeper lip of the roof membrane in the area of the adhesion flange of outlets should be compensated by a depression in the base or by use of an outlet of shallow flange thickness (e.g. foils).

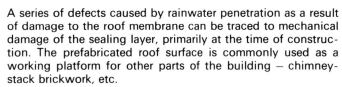

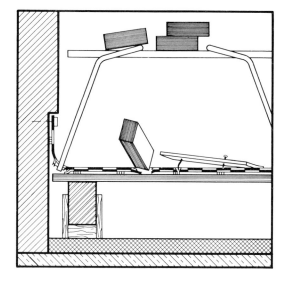

A series of defects caused by rainwater penetration as a result of damage to the roof membrane can be traced to mechanical damage of the sealing layer, primarily at the time of construction. The prefabricated roof surface is commonly used as a working platform for other parts of the building – chimney-stack brickwork, etc.

Bituminous as well as single-layer synthetic sealings may be damaged by:

— Excessive spot loading (e.g. pressure points caused by scaffolding poles).

— Loading (e.g. foot traffic) on built-up roofing layers that have lost their adhesion to the base.

— Working on the (unprotected) sealing layer.

— Dropping of sharp-edged objects (tools and construction materials).

— Movement of gravel because protective planking is not provided.

Points for consideration

— Sealing layers made of flexible bituminous felt or synthetic foils are sensitive to mechanical stresses. The risk of damage is increased where the sealing layer is laid on an uneven base or over gaps, such as the open expansion joints used with large slabs.

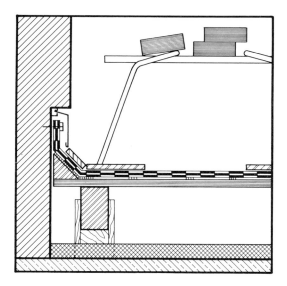

Recommendations for the avoidance of defects

● Protective layers made of gravel or chippings (compressed gravel layer), especially gravel fill, reduce the risk of damage to the sealing layers. Construction work on roof abutments, walls, etc., during which the sealing layer is often unprotected, should be carried out with appropriate precautions: boards prevent pressure points from scaffolding poles and damage due to movement of gravel or other work on the roof surface.

Double slab flat roof
Sequence of layers and individual layers

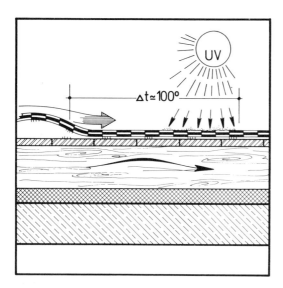

In unprotected or inadequately protected sealing layers damage through dampness, ponding and bubbles can occur as a result of climatic influences. In its unprotected state the roof membrane has been found to show various mechanical stresses (see B 1.1.11 – Stressing of sealing layer).

The surface protective layers applied to reduce such stresses may equally be defective: they may be blown away by wind, and thin reflective surface protection can become so dirty after a short period that the roof membrane is exposed to practically the same stresses as an unprotected sealing layer.

Points for consideration

– Unprotected sealing layers are exposed directly to the sun's rays. Increased brittleness and temperature differentials of up to 100°C on the roof surface are the consequences.

– The intense heat leads to large expansion which, in wide roof slabs, can overstress the roof membrane in the area of the joints. With lightweight double slab roofs this build-up of heat induces additional thermal loading.

– Ponding is particularly damaging to unprotected roofs because of the great differences in tension at the edges of puddles.

– Gravel fills give additional weather protection and also protection against mechanical damage due to applied load on the surface.

– Applied surface protective layers assist in reducing these stresses if distributed uniformly over the whole roof surface. Loose gravel fills must have a coarseness of $\geqslant 16$ mm dia; there is then no danger of their being blown away by the wind.

– Through weathering effects and dust deposits thin reflective layers rapidly lose their effectiveness, so that periodic renewal is necessary.

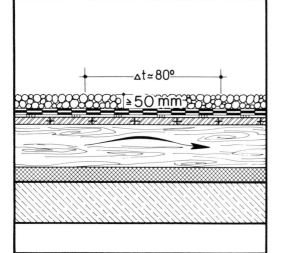

Recommendations for the avoidance of defects

● Every flat roof should have a surface protective layer. A gravel fill which can be used for a roof fall of $\leqslant 5°$ should have a coarseness of 16–32 mm dia and should be loosely and uniformly distributed. The loading resulting from this protective layer must be taken into account, with regard to depth, span and deflection, in the design and construction of the loadbearing structure.

● Roof surfaces which, because of a greater fall or lower load-bearing capacity of the upper slab, do not permit gravel fill to be used for the surface protective layer, should be protected by chippings, sanding or painting.

● Thin surface protective layers (reflective finishes, sanding, etc.) must be inspected at frequent intervals for their proper maintenance.

General texts and principles

Eichler, F.: Bauphysikalische Entwurfslehre, Band 2, 4. Auflage, Verlagsgesellschaft Rudolf Müller, Köln 1973.

Gösele, K.; Schüle, H.: Schall, Wärme, Feuchtigkeit, 2. Auflage, Bauverlag GmbH, Wiesbaden und Berlin 1972.

Henn, Walter: Das flache Dach, Verlag Georg D. W. Callwey, München 1967.

Hoch, Eberhard: Flachdächer – Flachdachschäden, Verlagsgesellschaft Rudolf Müller, Köln 1973.

Hoch, Eberhard: Kommentar Flachdächer, Verlagsgesellschaft Rudolf Müller, Köln 1971.

Jungnickel, Heinz u.a.: Abdichtungs- und Bedachungstechnik mit Kunststoffbahnen, Verlagsgesellschaft Rudolf Müller, Köln 1969.

Moritz, Karl: Flachdachhandbuch – Flache und flachgeneigte Dächer, 4. Auflage, Bauverlag, Wiesbaden und Berlin 1975.

Rick, Anton, W.: Das flache Dach, 5. Auflage, Straßenbau, Chemie und Technik, Verlagsgesellschaft mbH, Heidelberg 1966.

Seiffert, Karl: Richtig belüftete Flachdächer ohne Feuchtluftprobleme, Bauverlag GmbH, Wiesbaden und Berlin 1973.

Seiffert, Karl: Wasserdampfdiffussion im Bauwesen, 2. Auflage, Bauverlag GmbH, Wiesbaden und Berlin.

Zentralverband des Dachdeckerhandwerks: Richtlinien für die Ausführung von Flachdächern, Ausgabe Januar 1973, Helmut Gros Fachverlag, Berlin 1973.

DIN 4108: Wärmeschutz im Hochbau, Ausgabe 1969 mit den ergänzenden Bestimmungen, Okt. 1974.

DIN 18338: Dachdeckungs- und Dachabdichtungsarbeiten, August 1974.

DIN 18530 (Vornorm): Massive Deckenkonstruktionen für Dächer, Dez. 1974.

Sequence of layers and ventilation

Balkowski, F. D.: Die Ausbildung von zweischaligen Dächern. In: Das Baugewerbe, Heft 20/73, S. 81–84.

Buch, Werner: Das schwere und das leichte Flachdach. In: Deutsches Dachdeckerhandwerk (DDH), Heft 19/66, S. 1029–1035.

Buch, Werner: Grundlegende bauphysikalische Fragen des Flachdaches unter besonderer Berücksichtigung des Wasserdampfproblems. In: Bitumen, Teere, Asphalte, Peche..., Heft 10/71, S. 393–401.

Buch, Werner: Zur Theorie und Praxis flacher Dächer. In: Bauwelt, Heft 4/65, S. 90–95.

Caemmerer, W.: Berechnung der Wasserdampfdurchlässigkeit und Bemessung des Feuchtigkeitsschutzes von Bauteilen. In: Berichte aus der Bauforschung, Heft 51, Verlag Ernst & Sohn, Berlin 1968.

Cammerer, W. F.: Der Wärme- und Diffusionsschutz des Flachdachs nach dem heutigen Forschungsstand. In: Bitumen, Teere, Asphalte, Peche..., Heft 10/71, S. 402–412.

Haferland, Friedrich: Zur Bauphysik geneigter Dächer. In: Dachatlas, Geneigte Dächer, Institut für internationale Architektur – Dokumentation, München 1975.

Hauck, Werner: Belüftetes Flachdach, Punktförmige Wärmebrücken in der unteren Schale. In: Deutsches Architektenblatt (DAB), Heft 6/74, S. 400.

Heck, Friedrich: Kältebrücken in der Dämmschicht beim Flachdach. In: Deutsches Dachdeckerhandwerk (DDH), Heft 16/68, S. 1008–1011.

Hoch, Eberhard: Zweischalige Flachdachkonstruktionen – einst sichere Flachdächer, heute fragwürdige Funktionstypen. In: Das Dachdeckerhandwerk (DDH), Heft 11/75, S. 747–751.

Hoch, Eberhard: Das zweischalige Flachdach. In: Das Dachdeckerhandwerk (DDH), Heft 10/71, S. 632–635.

Götze, Heinz: Dämmschichten für Dächer und Außenwände. In: Das Bauzentrum, Heft 3/72, S. 65–85.

Kanis, Hellmut: Dächer sollten wieder geplant werden. In: Deutsche Bauzeitung, Heft 1/75, S. 44–45.

Koob, Hans K.: Wirtschaftliche Flachdachkonstruktionen aus Holz. In: Deutsche Bauzeitung, Heft 4/71, S. 693–698.

Probst, Raimund: Dächer – »kalt« oder »warm«? In: Baupraxis, Heft 5/69, S. 52–56.

Quassowski, Hans: Das Flachdach – immer noch Sorgenkind der Architekten? In: Der Architekt, Heft 9/73, S. 382–388.

Rick, Anton W.: Dachdurchlüftungen. In: Deutsches Dachdeckerhandwerk (DDH), Heft 4/67, S. 170–173.

Rick, Anton W.: Einige neue Beobachtungen zur Frage der Dachdurchlüftung. In: Das Dachdeckerhandwerk (DDH), Heft 1/70, S. 16–17.

Rick, Anton W.: Dampfsperre oder Lüftung? In: Bitumen, Teere, Asphalte, Peche..., Heft 10/67, S. 375–378.

Rick, Anton W.: Grenzen der Wärmedämmung von Flachdächern. In: Bitumen, Teere, Asphalte, Peche..., Heft 25/74, S. 360–363.

Schild, Erich: Bauschäden und Baufolgeschäden an Flachdächern. In: Industrie-Anzeiger, Heft 74/68, S. 20–22.

Seiffert, Karl: Dachdurchlüftungsfragen bei zweischaligen Wohnhaus-Flachdächern. In: Deutsche Bauzeitung, Heft 4/71, S. 677–678.

Seiffert, Karl: Gibt es grundsätzliche Vorteile beim einschaligen oder beim zweischaligen Dach? In: Bitumen, Teere, Asphalte, Peche..., Heft 18/67, S. 370–374.

Verhoeven, A. C.: Zur bauphysikalischen Beurteilung von Flachdachkonstruktionen. In: Detail, Heft 5/68, S. 889–894.

Zimmermann, Günther: Belüftetes Flachdach über Schwimmhalle, Feuchtigkeitsschäden infolge bauphysikalisch fehlerhafter Konstruktion. In: Deutsches Architektenblatt (DAB), Heft 22/73, S. 1859–1860.

Sealing layer and upper surface protection

Andernach, W.: Die Verklebung von Dachbahnen. In: Bitumen, Asphalte, Teere..., Heft 10/66, S. 378–380.

Götze, Heinz: Die lose verlegte Kunststoff-Dachhaut; Sonderdruck aus »Kunststoffe im Bau«, Themenheft 27.

Haage/Kramer: Neue Erkenntnisse über die Windbelastung auf Flachdächern. In: Das Dachdeckerhandwerk (DDH), Heft 14/74, S. 1446–1448.

Hafok, H.: Verlegung von Kunststoff – Dach – Dichtungsbahnen auf Basis Polyvinylchlorid (Weich – PVC). In: Deutsches Dachdeckerhandwerk (DDH), Heft 3/68, S. 105–112.

Haushofer, B.: Heißbitumenklebemassen aus der Sicht des Dachdeckermeisters. In: Deutsches Dachdeckerhandwerk (DDH), Heft 6/65, S. 244–247.

Haushofer, B.: Beitrag zur besseren Verklebung bituminöser Dachbahnen. In: Deutsches Dachdeckerhandwerk (DDH), Heft 16/67, S. 892–893.

Haushofer, B.: Bituminöse Schweißdachbahnen. In: Bitumen, Teere, Asphalte..., Heft 10/1969, S. 471–479.

Haushofer, B.: Bituminöse Dacheindeckungen auf geneigten Flächen. In: Das Dachdeckerhandwerk (DDH), Heft 13/72, S. 958–963.

Haushofer, B.: Lagen, Leitern, Löcher! In: Das Dachdeckerhandwerk (DDH), Heft 1/72, S. 22–25.

Holzapfel, W.: Dachbahnen. In: Das Dachdeckerhandwerk (DDH), Heft 2/72, S. 102–106.

Kakrow, H.: Dachabdichtungen aus der Sicht des Abdichtungsgewerbes. In: Bitumen, Teere, Asphalte...(BTAP), Heft 3/68, S. 82–86.

Kulhavy, A.: Oberflächenschutz bei verschiedenen Dachneigungen. In: Deutsches Architektenblatt (DAB), Heft 17/74, S. 1161–1162.

Müssig, Hans-Joachim; Hummel, Rudolf: Dachflächen grundsätzlich mit Gefälle ausbilden? In: Das Dachdeckerhandwerk (DDH), Heft 1/74, S. 70.

Müssig, Hans-Joachim; Hummel, Rudolf: Dachflächen grundsätzlich mit Gefälle ausbilden? Kommentierung der Flachdachrichtlinien. In: Das Dachdeckerhandwerk (DDH), Heft 1/74, S. 53–54.

Rick, Anton W.: Zur Frage der ebenen Dächer. In: Bitumen, Teere, Asphalte, Peche..., Heft 3/66, S. 112–114.

Rick, Anton W.: Verlegen von Deckungen und Dichtungen auf feuchten Flächen bei kaltem Wetter. In: Das Dachdeckerhandwerk (DDH), Heft 9/69, S. 520–523.

Rick, Anton W.: Vollaufklebung, Punktklebung oder loses Auflegen. In: Das Dachdeckerhandwerk (DDH), Heft 23/24/69, S. 1540–1542.

Rick, Anton W.: Warum Dachabkiesungen? In: Bitumen, Teere, Asphalte, Peche..., Heft 20/69, S. 473–474.

Problem: Connection to vertical abutments

Most double slab flat roofs have parapets and upstands formed from the roof structure, or are directly adjacent to higher neighbouring buildings. At these changes of plane the roof membrane must be connected to the vertical surface in such a way that the sealing function is unimpaired and the sealing edge is protected against damp penetration.

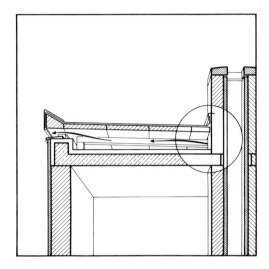

Empirical research investigations of building failure have shown that this junction of the sealing membrane to vertical surfaces is frequently damaged and presents a problem on the double slab flat roof.

Analysis of the damage shows a high incidence of typical, frequently recurring defects. These relate to faulty planning, to incorrect execution, but in the main to the poor quality of design of these points of detail, deriving from lack of knowledge of structures, of stresses and of correct use of materials.

It is possible to avoid many of the defects described in the following pages, and the faults which caused them, by following the recommended practice.

Double slab flat roof

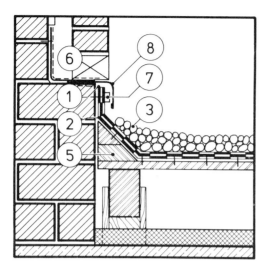

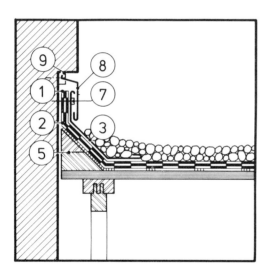

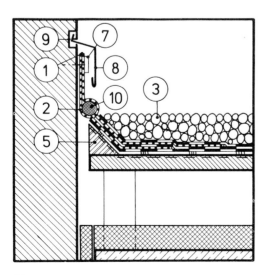

1 The roof membrane, at its connection to the vertical wall, must be raised above the highest level of water that may be expected in bad weather conditions. The height of the termination should be ⩾ 150 mm above the surface of the roof – gravel fill, compressed gravel layer (see B 2.1.2).

2 The termination and connection of the sealing layer to the vertical wall should be formed by the roof membrane itself or by using connecting foils which are materially similar in behaviour but better suited to the extreme stresses. A frictional connection of the pliable roof membrane with metal profiles or sheet metal should be avoided (see B 2.1.4).

3 The raised roof membrane must be protected from extreme effects of the external climate and from mechanical stresses (see B 2.1.4).

4 If the use of rigid metal sheets cannot be avoided, then these should be provided with sliding plates at adequate centres (approx. every 3 m) (see B 2.1.4).

5 The transition between the horizontal roof surface and the termination of the roof membrane on the vertical wall should be as gradual as possible, avoiding acute angles at the change of plane (see B 2.1.5).

6 To obtain a watertight roof connection the vertical structural surface must be waterproof. Facing slabs, claddings or any other non-watertight material must precede the sealing layer (see B 2.1.6).

7 In order to avoid slipping of the angled sealing layer, this should be fixed to the vertical surface uniformly along its whole length (see B 2.1.3).

8 The termination of the sealing layer must be protected at the connection point to the vertical surface from the onrush of water running down the face of the wall (see B 2.1.3).

9 Overhanging overflashings, which should fulfil sealing (water-diversion) functions, should be applied so as to move freely and should be subdivided by sliding seams (see B 2.1.3).

10 If movements are expected between the wall and the sealing roof base, the sealing must not be fastened to the vertical surface unless it is in a position to absorb the resulting deformations (see B 2.1.3).

Double slab flat roof
Connection to vertical abutments

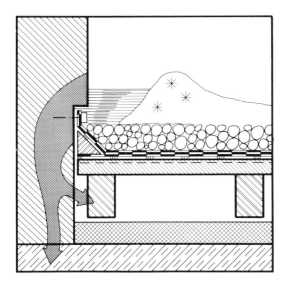

Damage resulting from dampness of the roof cross-section and the adjacent internal rooms occurs in double slab roofs where, in the area of connection to vertical surfaces, the roof membrane is not taken high enough above the surface of the roof structure. On flat roofs with minimum or no fall, and at depressions on the roof surface, the angled edge of the sealing layer may be submerged in water through the formation of puddles, or because of wind retention of melting snow.

Similarly, the vertical surface (e.g. wall surface) can become saturated to the inner wall surface by locally retained water or water spray.

Points for consideration

– It is impossible in practice for the edge of the sealing layer in connection with the vertical wall surface to be made watertight and it must therefore be given added protection.

– In the region of its junction with the roof the vertical wall surface is exposed to retained water to a degree comparable to that found in the foundation area of basements. Appropriate measures of protection against damp are required here.

Recommendations for the avoidance of defects

● At its junction with vertical parts of the structure, the roof membrane must be raised above the highest level at which water may be expected to enter in bad weather conditions. The ultimate height of the sealing membrane should be ⩾ 150 mm above the surface of the roof – gravel fill, compressed gravel layer, roof membrane. In extreme circumstances a greater height may be required, e.g. in the event of hollow formations in the roof surface or in tank roofs.

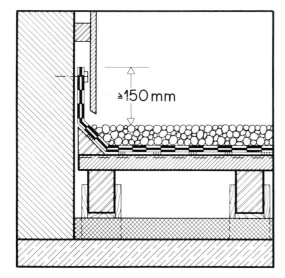

⩾150 mm

Double slab flat roof
Connection to vertical abutments

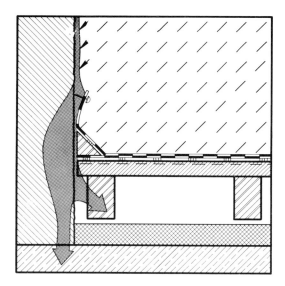

Rainwater may penetrate behind the edge of the raised sealing layer at its connection to the vertical surface (e.g. wall). This is the result of (a) inadequate protection to the ends of the sealing layer, which are fixed to the face of the abutting wall, or (b) vertical sealing layers which are inadequately fixed to the structural surface, being only spot fixed or glued on, and have come loose. Similar problems occur where metal flashings, provided to protect against damp, have worked loose from the joint or pulled apart at the soldered seam, leaving the edge of the sealing layer exposed to water.

Points for consideration

— The higher a vertical wall surface and the more freely it is exposed to driving rain, the heavier the water run-off from its surface. Every junction of the sealing layer which protrudes from this surface presents an obstacle to the running water and is strongly attacked by wind and water.

— The surface of the abutment wall is usually not completely uniform so that, for example, a pressed metal flashing finds no even and continuous bearing surface. Sealing by cement has very limited durability.

— The built-up roof membrane is subjected to strong thermal influences in the vertical plane, which can adversely affect its stability and that of the adhesion layer or its fixings and, as previously mentioned, pressed metal flashings are exposed to strong thermal deformations.

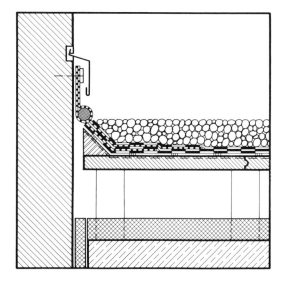

Recommendations for the avoidance of defects

● The edge of the sealing layer must be protected from penetration by running water at the junction of the angled sealing layer with the wall.

● In order to prevent the angled sealing layer from slipping, it should be fixed continuously along the whole length of its upper edge.

● If movements are expected between the wall and the flat roof – due to variable settlements – then the sealing layer should not be fastened to the wall unless it is in a position to absorb the resulting deformations.

● Metal flashings intended to fulfil a sealing (water-diversion) function should be free to move and should be subdivided by movement joints.

Double slab flat roof
Connection to vertical abutments

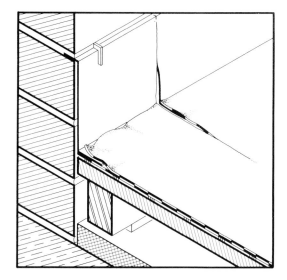

Where sheet metal of one-part metal profiles are used to form the junction of the sealing layer and the vertical structural wall, leakages are frequently observed. Long lengths of sheet metal without provision for expansion have been found to exhibit torn soldered seams and warping. If the metal adheres firmly to the roof skin, then this cracks, especially at the soldered seam. On unprotected areas of adhesion the built-up roof membrane separates from the adhesion surface. Alternatively, the connection of roof membrane and metal profile loosens in places, or along the whole length of the adhesion surface.

Points for consideration

— The use of sheet-metal angles necessitates the permanent, watertight connection of two very different structural materials by means of an adhesive.

— The vertical faces of the sealing layer are exposed to climatic influences and especially to changes in temperature. Depending on the material used, various changes in form or length are generated which stress excessively the roof membrane or other frictionally connected structural parts, leading to failure by compression (warping, folds) and tension (cracks).

— These stresses can be reduced by dividing the effective expansion lengths by means of movement joints. However, such provision for expansion is expensive and may give rise to new defects; wherever possible, firmly adhering wall connection sheets should be avoided.

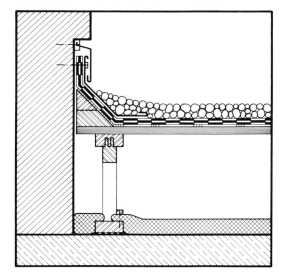

Recommendations for the avoidance of defects

● The fillet cover and junction of the sealing layer to the vertical abutment wall should be formed from the roof membrane itself or from connecting foils of similar material behaviour but better able to absorb extreme stresses. A frictional connection of the roof membrane with metal profiles or sheet metal should be avoided.

● The vertical roof membrane must be protected from the extreme effects of the climate and from mechanical stresses in the area of the angle. This is achieved by the application of cover flashings made of sheet metal. Sheet-metal profiles used as flashings must therefore move freely (i.e. not be firmly built in) and must be adequately subdivided by expansion joints in the form of overlap joints (with titanium zinc sheet, at least every 5–6 m).

● If the use of wall connection plates cannot be avoided, then these should be provided with movement joints at adequate intervals (with zinc sheets, at least every 3 m). The metal sheet, after removal of damp and dirt, should adhere for a width of ⩾ 120 mm to the multi-layer roof membrane. On inclined and vertical surfaces, special adhesives must be used.

Double slab flat roof
Connection to vertical abutments

In the angle of a roof where the built-up membrane of soft connecting foils – e.g. synthetic foils – changes plane, penetration has been found to result from mechanical damage: the pliable vertical and horizontal layers deflect, forming a right angle, and water leaks in where there is inadequate support at the junction.

Points for consideration

– The roof membrane is exposed to unforeseen mechanical stresses – e.g. foot traffic. Despite care, point loading can occur through a single gravel granule being subjected to extra loading; this can lead to the puncturing of the thin single-layer sealing membrane.

– The susceptibility of the sealing layer to mechanical damage is increased where it is unsupported.

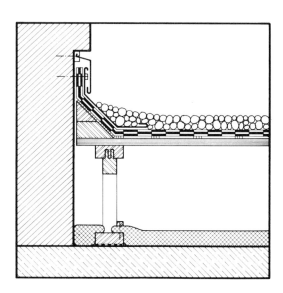

Recommendations for the avoidance of defects

● The transition between the roof surface and the abutment wall should, if possible, be designed so that a sharp angle at their junction is avoided. A gradual transition must be allowed for in the sealing layer, e.g. by the use of triangular fillets made of impregnated timber sections, so that a continuous sealing base is maintained.

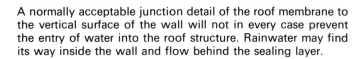

Double slab flat roof
Connection to vertical abutments

A normally acceptable junction detail of the roof membrane to the vertical surface of the wall will not in every case prevent the entry of water into the roof structure. Rainwater may find its way inside the wall and flow behind the sealing layer.

Points for consideration

– Protection against damp to outside walls, especially those constructed of more than one material, is not necessarily confined to the outer surface of the wall; the whole thickness of the wall must receive protection.
– Chimney-stacks are subject to considerable changes in temperature and consequently to deformation which may lead to cracks in the chimney coping or in the outer wall. Simultaneously, they are exposed to weathering effects and the whole outer wall may be saturated.

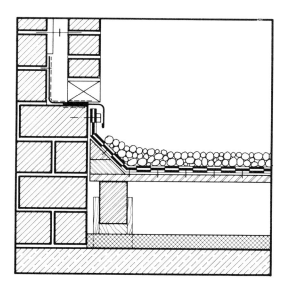

Recommendations for the avoidance of defects

● The satisfactory functioning of the roof connection presupposes that the surface of the wall is watertight. If this is not so, then the roof sealing membrane must be carried behind the permeable structural member in order to maintain a continuous protective skin to withstand damp penetration.

Problem: Edge of structural member

The structural design of roof edges is absolutely dependent on the fall and the method of drainage. Double slab flat roofs can be drained towards the perimeter by means of gutters or inwards by means of individual outlets. Accordingly, some roofs terminate at gutters and others have angled roof edges.

Particularly, double slab flat roofs require, in addition to an edge formation of the sealing layer appropriate to the construction, an effective arrangement of gutters or outlets together with a lateral protective covering (fascia) to the air space. These questions are dealt with in B 1.1.6.

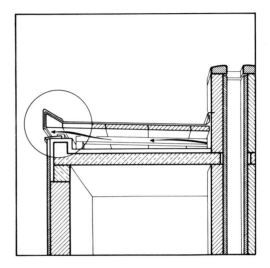

Sealing defects to roof edges with gutters were not found among the damaged double slab roof structures which were investigated. The predominant defect here – nearly as many as with single slab flat roofs – is in the roof form with internal drainage. Built-up angled roof edges present many difficult problems: over 40% of damage to points of detail was found to occur here. In some cases too little or no edge fall of the roof membrane, which is important in roofs planned without fall, was provided to the roof membrane; in others, the watertight edge formation of the sealing layer was defective.

In the following pages these defects are illustrated and analysed in more detail and recommendations for their avoidance are given.

Double slab flat roof

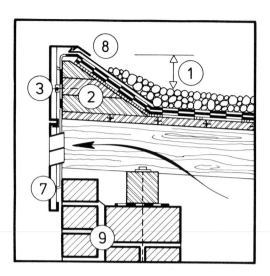

1 The edge of a roof drained inwardly must be raised above the roof drainage plane. The sealing layer must be raised $\geqslant 100$ mm above the upper surface of the roof – gravel fill, compressed gravel layer or other roof membrane (see B 2.2.2).

2 The roof edge should have an effective fall to the roof surface by means of a built-up angle of $\geqslant 30°$ (see B 2.2.2).

3 At the edge, the built-up roof membrane should be guided over an angle preformed in the sealing base to the required height of the slope, externally carried over the edge and fixed by nailing or clamp rails to the vertical face of the roof behind the applied fascia (see B 2.2.3).

4 A frictional connection of the pliable roof membrane with cover trims made of metal, asbestos cement or synthetic material should be avoided (see B 2.2.5).

5 If the sealing layer is attached to a roof edge cover trim, then this must securely anchor the roof membrane; at the same time, however, transfer of the deflections of the edge trim to the fastened roof membrane must be avoided (see B 2.2.4).

6 The roof membrane must be laid on a secure angled slope until it is fastened in the edge cover trim (see B 2.2.4).

7 The fixing of the roof edge trim must be secured to the supporting members of the roof edge, so that when load is applied there is no deflection with culminating stress effect on the roof membrane and the fascia, which is at a distance of $\leqslant 20$ mm in front of the external wall surface (see B 2.2.4).

8 The angled roof edge should be protected by loose fill gravel which is not connected with the sealing layer (see B 2.2.3).

9 Support walls terminating under the roof surface are preferable to a parapet formation (see B 2.2.6).

10 If parapets are provided, expansion joints must be arranged at centres of $\leqslant 5$ m. These expansion joints should be carefully sealed (see B 2.2.6).

11 The junction of the roof membrane with the parapet should be designed to accommodate the expected movements of the parapet section, especially in the area of the joints, otherwise cracking or separation of the roof membrane or connection profile will occur (see B 2.2.6).

12 Thermal bridges in the area of the roof edge should be avoided by using thermal insulation material or internal insulation; with slight angles of inclination, complete cladding with thermal insulation material is required (see B 2.2.6).

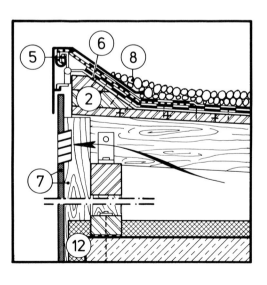

Double slab flat roof
Edge of structural member

With double slab flat roofs, in particular those with lightweight upper slabs, frequently the roof membrane at the roof edge has little or no slope from the roof plane. In roofs with little or no fall, wind banking or defective fall leads to rainwater spilling over the roof edge and results in the soaking and soiling of the external walls. The water also soaks through the roof structure, with consequent damage and rotting of timber sections and dampness inside the building.

Such damage affects lightweight timber roofs with cantilevered roof edges. Owing to sagging, these surfaces no longer drain inwards, so that with low roof slopes the rainwater percolates over the edge (see B 1.1.7 – Deformation of upper slab).

If the roof surface is loosely covered with a protective layer of very fine gravel, this loose gravel layer will be blown by the wind over the roof edge. Such falling gravel has been known to cause damage.

Points for consideration

— The roof edge may be exposed to increased water pressure through a defective fall or one induced by distortion of parts of the roof. This applies particularly to lightweight roofs designed with no fall.

— Increased loading can be created by wind banking at the roof edge increasing the depth of water on certain areas of the roof.

— Equally, a backwash can be caused by blocked outlets at roof drainage points, resulting in a build-up of water. Low roof upstands are then ineffective in stemming the flow of water.

— The stressing of the sealing edge can be reduced if the rainwater can be speedily diverted by an effective fall from the vulnerable edge to the roof surface.

— Loose gravel of low weight and < 16 mm dia is easily shifted by wind and may be blown over roof edges with too low an edge upstand.

Recommendations for the avoidance of defects

● The edge of inwardly drained roofs must be raised above the level of water that may be expected in severe weather conditions. The sealing layer must lie ⩾ 100 mm above the surface of the roof – gravel fill, compressed gravel layer, or any other roof membrane. In special circumstances a higher edge upstand may be required.

● The roof edge should have an effective angle of fall (⩾ 30°) to the roof surface, by means of a built-up edge angle fillet in the sealing layer.

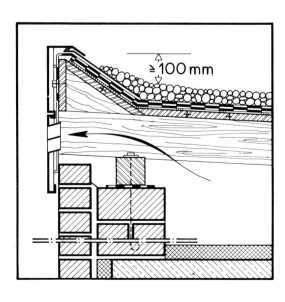

≥100 mm

Double slab flat roof
Edge of structural member

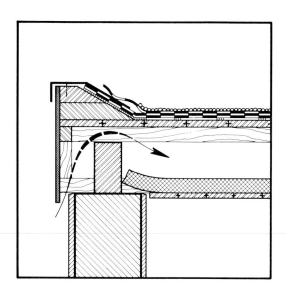

A large number of defects to roof edges relate to the edge formation of the sealing layer. Trims made of metal, asbestos cement, or synthetic material fixed with adhesive to the roof membrane lead to the loosening of the adhesion; sometimes the strip comes apart from the adhesion flange (see B 2.2.5). Sealing layers laid on angled or vertical surfaces slip and develop folds. If the end of the sealing layer is not clamped far enough into the roof edge trim it comes loose.

These sealing defects cause water penetration of the roof structure and of the rooms inside. If materials used for the upper slab or thermal insulation layer are not stable in wet conditions, severe damage to the whole roof structure results (see B 1.1.7).

Points for consideration

- Joints of the sealing layer are vulnerable because of the necessary adhesion and the easily damaged lateral edges of the strips and foils.
- The sealing layer in the area of the roof edge upstand is greatly exposed to climatic influences. The junctions of materials of different characteristics are particularly stressed.
- Because of the danger of slipping, permanent protection of the angled sealing layer cannot be achieved by means of gravel. Protective coatings have only a limited life and require periodic renewal.

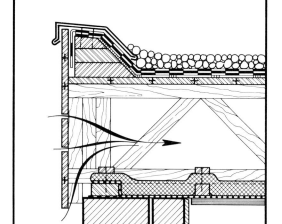

Recommendations for the avoidance of defects

● At the roof edge the roof sealing membrane should be laid over an angled upstand (made of a glued timber section) preformed in the sealing base (upper slab) to a distance of ⩾ 100 mm up the slope, being nailed at the top of the upstand or on the vertical face.

● The angled roof edge should be protected by interlocked, in some cases pre-sprung, trims laid loosely on the sealing membrane and protecting the junction.

Double slab flat roof
Edge of structural member

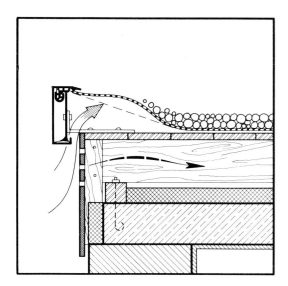

Various types of defect occur in roof edges where the roof membrane is clamped into a multi-part edge trim. The edge connection may be cut too short and therefore inadequately secured, so that after a time it loosens. Where edge membrane connections have been laid under stress and are unsupported in the upstand area, damage results from traffic on the roof surface. If the multi-part edge trim is covered with adhesive, cracks appear in the clamped roof membrane, primarily in the joint area of the clamp trims. Storm damage caused by wind uplift may occur on roof edges with loose-laid edge foils and fascias mounted at considerable distances from the wall surface.

Points for consideration

— The strengthening of the edge of the sealing layer by metal cover trims requires a form of construction that anchors the roof membrane securely, but which avoids transference of the deformation of the roof edge sections to the roof membrane. This can be achieved by means of multi-part cover trim with clamping fixings.

— Contamination of the cleat section by adhesive, etc., as well as inaccurate erection, can lead to unintended frictional connections of the roof membrane and the cover trim and cause cracking.

— If the edge upstand is formed by means of the metal section alone, the roof membrane is unsupported. The stretched and unsupported sealing membrane is particularly vulnerable to mechanical stresses.

— The cleats and cover trims are subjected to other external stresses under certain circumstances, e.g. leaning of ladders. These trims, if of slender section or if not fixed to the structural part of the roof edge, may be disfigured so that the roof membrane contained in the trim is unduly stressed.

— Wind stress to the roof is increased if, added to the suction effect on the roof surface, there is a considerable space between the fascia and the wall. This open area allows the wind pressure to build up directly below the sealing membrane, which is unsupported, and causes it to crack.

Recommendations for the avoidance of defects

● If the sealing layer is connected to a roof edge cover trim, then this must securely anchor the roof membrane. However, the transfer of deformations from the edge trim to the clamped roof membrane must be avoided under normal building conditions. This can be achieved by the use of multi-part, detachable cover trims.

● The roof membrane must be raised to be clamped into the edge trim on a firm upstand which serves as a sealing base.

● The securing of the roof edge trim to the loadbearing part of the roof edge must be such that, when stresses take effect, no deformation results that will affect the clamped roof membrane. The ventilated openings should be at a distance of $\leqslant 20$ mm in front of the external wall surface.

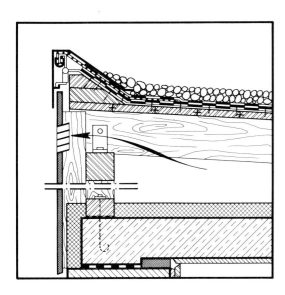

Double slab flat roof
Edge of structural member

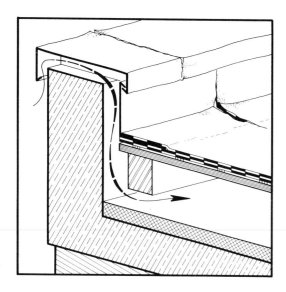

In double slab roofs with sealing edges formed of metal trims or plates frictionally connected to the sealing layer or clad with wide metal sheets, leakages occur leading to soaking of the roof structure and the rooms below.

The following specific defects have been observed.

(i) In a built-up roof membrane where the sealing layer adheres to the metal cover section, cracks appear in the roof membrane near the open joints or at the soldered seams of the metal trims.

(ii) The adhesive between the roof membrane and the metal surfaces deteriorates in places, sometimes along its whole length.

(iii) At inclined unprotected adhesion points, the sealing membrane separates from the sheet metal and leaks appear, particularly where the area of adhesion is too small.

(iv) In areas where large sheet-metal copings are used at the roof edge (e.g. parapet coping), warping, buckling and cracked soldered seams occur through which rainwater penetrates the roof structure and the external walls.

Points for consideration

— To seal the roof edge upstand successfully, using cover trims made of metal, asbestos cement or synthetic material, requires the permanent connection of two very different construction materials by means of an adhesive.

— The roof edge is exposed, without protection, to climatic effects and in particular to changes in temperature. According to the material used, considerable linear expansion and changes in form can be generated, which excessively stress the roof membrane or other frictionally connected components and lead to failure by warping and cracking. By subdivision of the effective linear expansion and provision of expansion joints, these deformations are reduced; however, the construction of such provision for expansion is expensive and conceals new weaknesses.

— Strong sunlight on the often unprotected adhesion flanges leads to softening of the adhesive so that the frictional connection is extensively destroyed. Under tension, displacement, loosening or slipping of the sealing layer from the adhesion flange can occur.

— Dirt, damp and low temperatures prevent a watertight connection of the roofing panels to the metal adhesion flanges.

Recommendations for the avoidance of defects

● The sealing edge should be formed from the roof membrane itself or by using a connecting foil made of a material of similar characteristics. A frictional connection of the pliable roof membrane with trims made of metal, asbestos cement or synthetic material should be avoided.

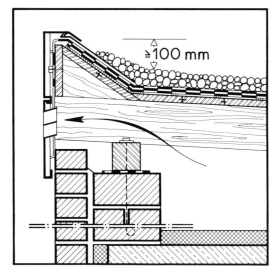

Double slab flat roof
Edge of structural member

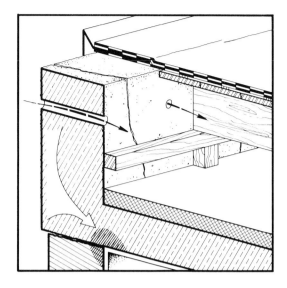

At the reinforced concrete upstand or parapet of solid roofing slabs on double slab ventilated roofs, damage in the form of vertical cracks at intervals of a few metres frequently occurs. This leads to soaking of the under-side of the roof. In addition, with this form of roof design, under-dimensioned or complicated ventilation openings are often provided, so that the satisfactory functioning of the whole roof is destroyed.

Where upstands (parapets) are constructed above roof level, wide metal copings warp. If the roof membrane is fixed at a considerable distance above the angle fillet, overstressing occurs and cracks appear in the sealing layer at the outer corners (deflection points). Similarly, sealing layers which are either inadequately fixed or unstable slip and develop folds, ultimately working loose from their base.

Points for consideration

— With double slab roofs, upstands are used as roof beam supports and lateral termination of the roof spaces. Where the depth of the roof construction is considerable (duct space, fall > 5°), the construction of upstands or parapets can be practical despite the problems described.

— Because of its more protected position and the omission of complicated roof membrane junctions, a wall terminating under the roof surface is less liable to damage than one which extends above the roof surface as a parapet. The arrangement of ventilation openings is also simpler than with parapets.

— With edge upstands designed as parapets, the large thermal linear expansions are accommodated by closely spaced expansion joints or by strong additional reinforcement.

— Large sheet cladding which is necessary for sealing must be subdivided at frequent intervals by sliding coverjoints, because of expansion due to sun exposure.

— Sealing layers laid unprotected over parapets are subject to stress from ultra-violet rays as well as from thermal expansion of the parapet. The application of gravel or protective paint gives slight protection but only for a limited period.

— Reinforced concrete parapets which are firmly fixed or part of the roof slab form a thermal bridge which can be avoided only by internal insulation — structurally this is not recommended.

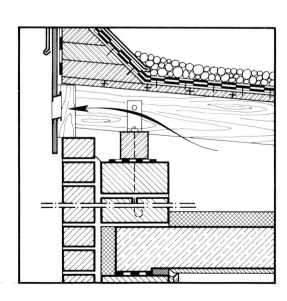

Recommendations for the avoidance of defects

● Walls terminating under the roof surface are preferable to a parapet.

● If a parapet is built it must have expansion joints ⩾ 20 mm wide at centres of ⩽ 5 m. These expansion joints should be carefully sealed.

● The junction of the roof membrane with the parapet must be designed so that the anticipated movements of the parapet sections (especially in the area of the joints) do not lead to fracture or loosening of the roof membrane or at the connecting cover trim.

● Thermal bridges in the area of the parapet should be avoided by use of insulation material; shallow upstands should be completely covered by the thermal insulation material.

Problem: Bearing surface and expansion joints

In the ventilated double slab roof the bearing area presents a problem in lightweight timber structures as well as in heavy reinforced concrete structures. With solid roof slabs the changes in form present the main problem; with lightweight structures it is the weathering stresses.

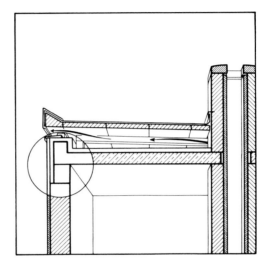

At the edge of the ventilated roof near the bearing surface, the problems of termination of the sealing layer and the arrangement of the ventilation openings have to be solved simultaneously. The bearing and edge area require early detail designing, especially in the double slab roof.

The main defects and the structural recommendations are illustrated in the following pages.

Double slab flat roof

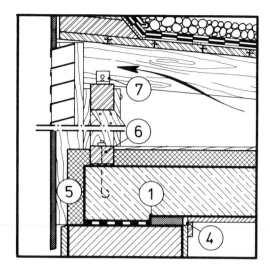

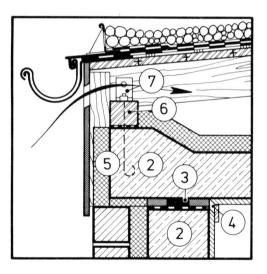

1 If the roof bearing surface is designed as a rigid connection, separating foils should be laid in the bearing joint and a frictional connection in the inner third of the bearing surface avoided (e.g. by lining with a thermal insulating strip approx. 10 mm thick) (see B 2.3.2).

2 Solid roofing slabs of double slab roofs on loadbearing brickwork should be provided with expansion joints where structural members are longer than 12 m. They should be displaceably mounted if the linear expansion is slight or has not been proved by calculation. In addition to the dowel reinforcement of the roof slab there should be a dowel connector over all loadbearing walls and functional slip foils or bearings in the bearing joint. Non-loadbearing walls should be separated from the roof slabs (see B 2.3.3).

3 The surfaces of the slip joint should be smooth; in reinforced brickwork a cement mortar smooth coating $\geqslant 30$ mm thick should be applied as packing. The slip foil used should be capable of compensating for the remaining unevennesses by the use of flexible strips (see B 2.3.3).

4 Ceiling and wall plaster should be separated in the region of the slip joint. The external joint should be covered in a manner that prevents penetration by rainwater (see B 2.3.2 and 2.3.3).

5 The external vertical face of the roof edge, and if necessary the upstands, should be protected by external quality thermal insulation of $\geqslant 1 \cdot 3$ m² K/W. This should be continuous with the thermal insulation layer of the roof structure (see B 2.3.2 and 2.3.3).

6 Lightweight timber roof structures should be secured to the bearing walls by bolt and screw connections. The timber rafters of lightweight roofs should be supported on wall plates. The wall plates should be anchored into the brickwork at a distance of $\leqslant 2$ m (see B 2.3.4).

7 If the rafters are nailed to the wall plate, at least every third rafter is to be bolted to the wall plate by means of mild steel angles (see B 2.3.4).

Double slab flat roof
Bearing surface and expansion joints

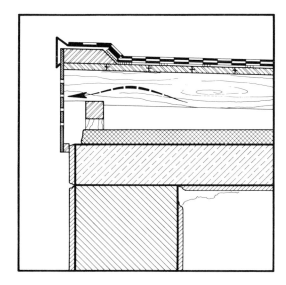

In ventilated double slab roofs with solid reinforced concrete roof slabs as the lower slab, despite shortness of the span (< approx. 12 m) and adequate thermal insulation on the upper face of the slab, horizontal cracks have been found to occur in the area of the bearing joint. These cracks are apparent in the ceiling and wall plaster. If the bearing area is not covered externally by a ventilated fascia, rainwater penetrates.

If the roof edge is not insulated, dampness and discoloration appear at the corners of the room inside.

Points for consideration

— The sagging of the reinforced concrete slabs produces distortions in the bearing surface which, with continuous bearing over the whole wall, can lead to eccentric loading and cause cracks in the bearing surface joint.

— A joint between the roof slabs and bearing surface brickwork is usually unavoidable because of sagging. Continuous plastering over the joint is not possible without the formation of cracks.

— The roof upstand already forms a thermal cold bridge because of its vertical cold surface. With the double slab roof this cooling surface is increased by upstands which form the lateral termination of the air space. The thermal insulation of the external face of the roof and its connection to the horizontal thermal insulation layer are important, if low surface temperatures and subsequent formation of condensation are to be avoided.

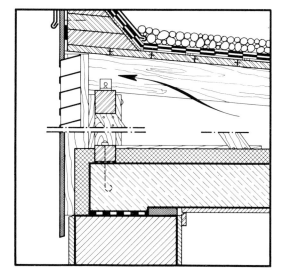

Recommendations for the avoidance of defects

● If roof bearing surfaces can be made rigid, so that the expected linear expansion does not lead to damage, then a separating foil or strip should be laid in the bearing joint (e.g. by inserting a thermal insulation strip approx. 10 mm thick) so that a frictional connection in the inner third of the bearing surface is prevented.

● The ceiling and wall plaster should be separated in the area of the bearing joint. The outside face of the joint should be covered with a plaster stop to prevent penetration by rainwater. The fascia enclosing the air space can be used as a ventilator.

● The external vertical surface of the roof edge should be protected with external quality thermal insulation with an insulating value of $\geqslant 1 \cdot 3$ m² K/W. This should be continuous with the thermal insulation layer of the roof structure.

Double slab flat roof
Bearing surface and expansion joints

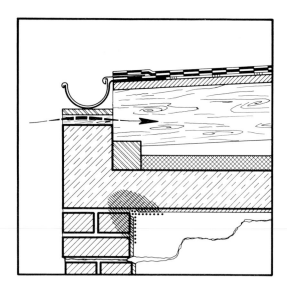

Horizontal and stepped cracks – especially at the corners of the building – may occur at the brickwork supports of solid roofing slabs on double slab roofs.

These defects may occur in slabs with adequate external insulation where the length of the structural member exceeds 12 m and the bearing is designed without dowel, slip joint or expansion joint.

If the roof edge, which is often designed with an upstand, is not insulated, water penetrates to the room below at the angle of ceiling and wall.

Points for consideration

– If linear expansion of solid roofs is prevented by sliding joints, pressure builds up which, in members longer than 12 m, leads to fracture of the bearing brickwork. In such cases, sliding joints and dowels must be provided at the loadbearing wall junction, unless the linear expansion over the bearing surfaces is reduced by an increase in thermal protection or by expansion joints.

– The construction of dowels with a sliding joint between roof and bearing surface can lead to a tension-free dispersal of the expansion, if synthetic joints with low frictional resistance are used.

– The functioning of sliding foils is very dependent upon the smoothness of the bearing surface. Unavoidable irregularities must therefore be corrected by increasing the thickness of the sliding foils and bearings.

– Roof coverings which are continuous over the slip joints cannot accommodate the movements which occur; they crack over the joint.

– Uninsulated vertical roof planes and upstands form thermal bridges which can lead to the formation of condensation on the inner surface.

Recommendations for the avoidance of defects

● Solid roof slabs of double slab roofs on loadbearing brickwork should be provided with expansion joints or be freely mounted when the structural members span more than 12 m. In addition to dowel reinforcement of the roof slabs there should be a dowel connector over all loadbearing walls and functional slip foils in the bearing joint. Non-loadbearing walls should be separated from the roof slab by a bituminous membrane.

● The adjoining faces of the slip joints should have even surfaces. The sliding membrane used should be capable of countering the remaining friction by means of additional flexible strips.

● Wall and ceiling plaster should be separated at the junction of the slip joint. The wall joint should be protected so that rain cannot penetrate.

● The vertical faces of the roof, fascia and upstands should have external thermal insulation of $\geqslant 1 \cdot 3$ m² K/W, which must be continuous with the thermal insulation layer of the roof structure.

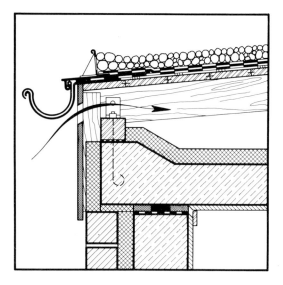

Double slab flat roof
Bearing surface and expansion joints

Double slab flat roof
Bearing surface and expansion joints

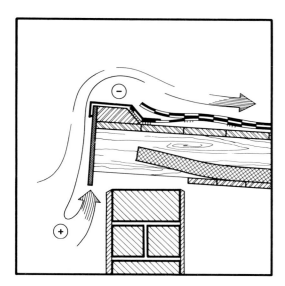

Where buildings have lightweight double slab roofs, or double slab roofs with a lightweight upper slab, it frequently occurs that the complete timber roof structure is lifted and destroyed in a storm.

Points for consideration

— Lightweight timber roof structures are unable to counteract wind forces by their dead weight. The anchoring of the individual parts of the structure to each other and to the loadbearing wall members is most important in lightweight roofs.

— Rafters laid directly on the loadbearing brickwork are difficult to anchor securely. Fixing onto a wall-plate which is itself anchored in the brickwork provides safer construction. Fixing by nails alone has proved to be inadequate in storm conditions.

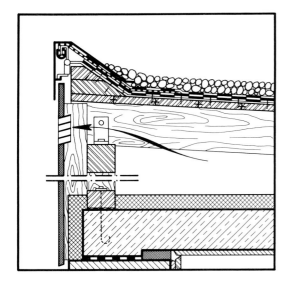

Recommendations for the avoidance of defects

● To be secure from storm damage, lightweight timber roof structures should be connected by bolt and mild steel angle connections to the bearing walls.

● The timber rafters of lightweight roofs should be secured to wall-plates. The wall-plates should be anchored at centres of $\leqslant$ 2 m.

● If the rafters are nailed to the wall-plates, at least every third rafter should be additionally joined to the wall-plate by means of m.s. angles and bolts.

Problem: Drainage

In the interest of ventilation, double slab roofs should have the minimum fall to the upper slab that provides, at low cost and maintenance, an even ceiling below. Rainwater can be drained away fairly easily by means of external gutters.

However, double slab roofs are frequently constructed with surfaces having slight fall or no fall, with gutters and/or outlets on the inner face of the roof where the drainage is of considerable significance. As the roof pitch is reduced, the requirements of waterproofing and evenness of the roof membrane become greater, since the dispersal of the rainwater is slower and the sealing layer is exposed to water for a longer period of time. Small unevennesses (caused by welts, for example) or slight sagging of the loadbearing structure can prevent the rainwater from flowing away easily.

In lightweight double slab flat roof structures, the additional dead load of standing water resulting from blocked outlets or drainpipes must be taken into account.

It is important at the building design stage to consider the discharge of rainwater from flat roofs and locate and size outlets, etc. The selection of outlet patterns, gullies and downpipes is related to the structure of the roof for falls and to their connection with the sealing layer or thermal insulation.

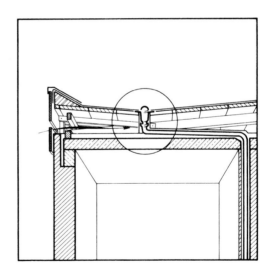

While external gutters appear to be less liable to damage – approximately 25% of all the double slab roofs investigated had external drainage on which no damage was reported, and are not included in the following analysis of defects – the incidence of damage to internally lying gutters and outlets is considerable.

Connection details, gutters, individual outlets, unfavourably situated drainage elements and gutters lying transversely to the direction of inner roof ventilation were found to be particularly vulnerable. Further details on these main areas of defects are given in the following pages.

Double slab flat roof

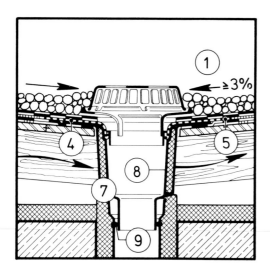

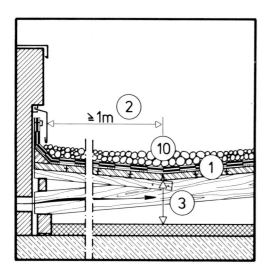

1 In principle, flat roof outlets should be located at the lowest part of the roof and drain the surrounding roof areas by means of an effective fall ($\geqslant 3\%$). Deflections, especially in light-weight structures and cantilevered slabs, must be taken into account (see B 2.4.2).

2 Individual outlets and gutters should be situated at a distance of $\geqslant 1$ m from roof abutments such as parapets, upstands, walls, etc. (see B 2.4.2).

3 If the installation of channel or box gutters lying at right angles to the direction of the inner roof ventilation cannot be avoided, care must be taken to provide sufficient depth of ventilation ($\geqslant 100$ mm) in the region of the shallow part of the roof (see B 2.4.2).

4 Flat roof rainwater outlets must be constructed in a manner that ensures perfect connection to the roof membrane. They should be provided with a sufficiently large, durable, but flexible, connection flange (see B 2.4.3).

5 Care should be taken to achieve good adhesion between the roof sealing membrane and the adhesion flange or connection foil of the outlet element. To allow for movement, the sealing layer should be secured only at the outer rim of the connecting flange, by means of adhesive (see B 2.4.3).

6 The increased thickness at the flanges of the outlet funnel, where the sealing layer is affixed, must be counteracted by a correspondingly reduced section of the loadbearing slab, or by using an outlet unit of thin flange section (e.g. foil) (see B 2.4.4).

7 Outlets and drainpipes should be lagged with thermal insulation material (see B 2.4.4).

8 In double slab roofs the construction of single-stage outlets, of which the flanges or connection foils can be bonded into the sealing panels on the upper slab, is adequate (see B 2.4.4).

9 As protection against water seepage at the joint, a preformed groove and gasket seal should be provided between the outlet elements (see B 2.4.4).

10 Box gutters should not be used because of their difficult and insecure connections. Channel gutters formed directly from the roof membrane are preferable (see B 2.4.4).

Double slab flat roof
Drainage

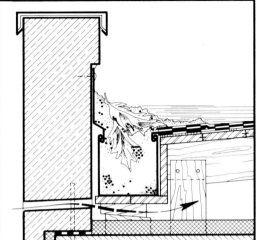

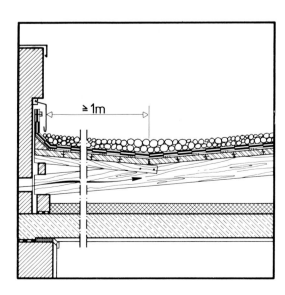

Flat roof outlets may develop defects of function after a short time, especially if they are unfavourably located on the roof surface to be drained. In outlets which lie very close to the roof abutment this is most noticeable: the connections to the sealing membrane are not sealed, or the outlets become blocked by leaves, dirt or snow, so that the water can no longer flow away but lies in puddles on the roof surface.

The formation of puddles with subsequent cracking may also be discovered in roof surfaces of uneven depth, in which the outlets lie in the higher part of the roof.

Where box gutters are installed, condensation damage is often found within the roof section.

Points for consideration

— Individual outlets and gutters can effectively drain rainwater only from that part of the roof surface where the direction of the fall is towards them. Standing water and consequent cracking in the area at the edge of the puddle must be avoided (see B 1.1.10 – Formation of ponding on the roof membrane).

— In the design of the drainage elements possible deflection should be taken into account, especially with lightweight structures and corbel or cantilevered slabs, to ensure that the outlets lie in the deepest areas of the roof.

— Outlets and gutters must be fastened to the roof membrane by means of connecting elements (flanges, connection foils) and therefore need sufficient room for their installation.

— If outlets and gutters are located close to roof abutments, parapets or upstands, leakage at the junction with the roof membrane is likely, as effective waterproofing is made difficult, if not impossible, by lack of working space. Additionally, there is the danger of ice forming at the outlet, so that the installation of heated drainage elements may become necessary at very exposed sites.

— Wind causes leaves, dirt and snow to accumulate, particularly in the corners and channels of roof surfaces, and these prevent drainage by obstructing the water flow to outlets and gutters.

— Box gutters, because of their depth, extend deep into the ventilation space. When located at right angles to the direction of ventilation, they prevent a regular exchange of air and condensation damage can result.

Recommendations for the avoidance of defects

● In principle, flat roof outlets must lie in the lowest areas of the roof surface, with an effective fall ($\geqslant 3\%$) of the drainage area.

● Individual outlets and gutters should be at a distance of $\geqslant 1$ m from roof upstands such as parapets, walls, etc.

● Where box gutters at right angles to the direction of ventilation cannot be avoided, care must be taken to allow adequate ventilation space ($\geqslant 100$ mm deep) even in the lowest area at the end of the fall.

Double slab flat roof
Drainage

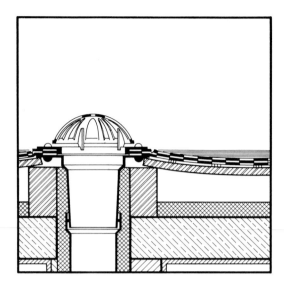

In drainage away from the edge of the roof, drainage channels, as well as individual outlets (gullies), have been found to show defects of construction at the connection to the sealing layer. Incorrect adhesion between the sealing membrane and preformed metal sections causes leakage through the roof structure and damages the ceiling below by partially loosening the plaster.

Points for consideration

— Because of their function and position in the lowest part of the roof surface, drainage channels are subjected to strong stresses from rainwater. Leakages therefore have a particularly damaging effect.

— Effective roof outlets close to, or even built into, parapets or roof edge upstands are very difficult to construct (see B 2.4.2 – Arrangement and position of drainage elements).

— Where there is a large variation of temperature between different materials (e.g. bituminous roof membrane – cast iron roof outlet) considerable stresses arise which must be balanced in the transition areas (e.g. by the use of flexible connecting layers).

— Overlaps and strengthening layers in the area of the outlet flange produce an increase in thickness at the outlet rim, impeding the flow of rainwater. This results in the formation of puddles.

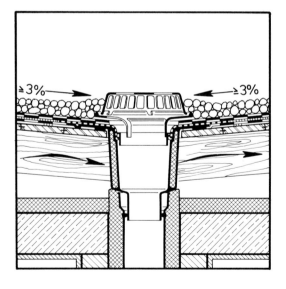

Recommendations for the avoidance of defects

● The flat roof outlet must be so constructed as to ensure a satisfactory connection to the roof membrane.

● Rigid connections without stress compensation should not be made. Outlets in flat roofs should be provided with an adequate flexible connecting flange in every drainage plane.

● Special care should be taken to provide good adhesion of the sealing layer to the appropriate adhesion flange. To allow for movement, the sealing layer should be secured only in the outer area of the connecting flange.

● Overlapping in the area of connection of the roof membrane with the drainage outlet results in increased thickness. This must be compensated by a reduction in the loadbearing slab thickness in the bearing area of the connection flanges. If this is not possible, drainage elements with thin flange sections (e.g. foils) should be used.

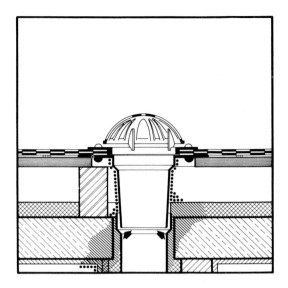

Damage to drainage elements caused by dampness occurs particularly where drainpipes, outlets or channels have been installed without thermal insulation lagging, and where incorrect outlets or defectively constructed channels or box gutters have been built into the structure. After the defective sealing connections already mentioned, these errors have been found to be the most frequent cause of failure in flat roof drainage.

Points for consideration

— Roof outlets and the connected drainpipes present problems of thermal technology as well as of acoustics. Where uninsulated outlets and pipes are installed, condensation forms on the under-side of the roof and on the drainpipes. This results in dampness to the structure (thermal insulation, concrete roof and walls).

— Where drainpipes are blocked or outlets are too narrow in cross-section, water must not be able to penetrate the roof structure through backwash.

— Drainage channels and box gutters of preformed sheet metal present an exceptional hazard because of the arrangement of sliding seams and folds at their junctions.

Recommendations for the avoidance of defects

● Outlets and drainpipes should have thermal insulation lagging. The installation of heated roof gullies in frost-prone areas provides an additional safeguard against frost damage to a flat roof, as these parts can be kept free of ice (not normally installed in the British Isles, but common on the continent of Europe).

● In double slab roofs the installation of single-stage outlets, of which the flanges or connection foils can be bonded into the sealing panels on the upper slab, is adequate.

● For protection against water seepage at the joint, a preformed groove and gasket seal should be provided between the outlet elements.

● Box gutters should not be used, because of their difficult and insecure connections. Trough or channel gutters which can be formed directly from the roof membrane are recommended.

Double slab flat roof
Points of detail

General texts and principles

Eichler, Friedrich: Bauphysikalische Entwurfslehre, Band 2, 4. Auflage, Verlagsgesellschaft Rudolf Müller, Köln 1973.

Hoch, Eberhard: Flachdächer – Flachdachschäden, Verlagsgesellschaft Rudolf Müller, Köln 1973.

Jungnickel, Heinz u.a.: Abdichtungs- und Bedachungstechnik mit Kunststoffbahnen, Verlagsgesellschaft Rudolf Müller, Köln 1969.

Meyer-Bohe, Walter: Dächer, Verlagsanstalt Alexander Koch GmbH, Stuttgart 1972.

Moritz, Karl: Flachdachhandbuch – flache und flachgeneigte Dächer, 4. Auflage, Bauverlag Wiesbaden und Berlin 1975.

Schild, E.; Oswald, R.; Rogier, D.: Bauschäden im Wohnungsbau Teil II – Bauschäden an Dächern, Dachterrassen, Balkonen – Ergebnisse einer Umfrage unter Bausachverständigen, Verlag für Wirtschaft und Verwaltung – Hubert Wingen, Essen 1975.

Zentralverband des Dachdeckerhandwerks: Richtlinien für die Ausführung von Flachdächern, Ausgabe Januar 1973, Helmut Gros Fachverlag, Berlin 1973.

DIN 1045 – Beton- und Stahlbetonbau, Januar 1972.

DIN 1053 – Mauerwerke, Berechnung und Ausführung, Blatt 1, November 1974.

DIN 1055 – Ergänzende Bestimmungen zu Blatt 4, Lastannahmen im Hochbau, Verkehrslasten, Windlasten; Ausgabe Juni 1939, März 1969.

DIN 4122 – Abdichtung von Bauwerken gegen nicht drückendes Oberflächenwasser und Sickerwasser mit bituminösen Stoffen, Metallbändern und Kunststofffolien, Richtlinien, Juli 1968.

DIN 18530 – (Vornorm) Massive Deckenkonstruktion für Dächer, Dezember 1974.

Connection to vertical abutments

Buch, Werner: Wärmedämmung von Dächern mit Hartschaum aus Styropor, Herausgeber: Informationszentrum Styropor.

Haushofer, Bert: Ein Flachdach ist stets so gut wie seine Anschlüsse. In: Deutsches Dachdeckerhandwerk (DDH), Heft 10/68, S. 571–575.

Hummel, J.: Abschlüsse und Anschlüsse im flachen Dach. In: Deutsches Dachdeckerhandwerk (DDH), Heft 14/69, S. 870–871.

Kakrow, K. H.: Detailausbildung in der Flachdachabdichtung. In: Deutsches Dachdeckerhandwerk (DDH), Heft 1/68, S. 25–30.

Schaupp, Wilhelm: Das bituminöse Dach ist stets so gut oder so schlecht wie seine Anschlüsse. In: Deutsches Dachdeckerhandwerk (DDH), Heft 19/69, S. 1267–1282.

Schlenker, Herbert: Die Fachkunde der Bauklempnerei, A. W. Geutner Verlag, Stuttgart 1971.

Edge of structural member

Haushofer, Bert: Ein Flachdach ist stets so gut wie seine Anschlüsse. In: Deutsches Dachdeckerhandwerk (DDH), Heft 10/68, S. 571–575.

Hummel, J.: Abschlüsse und Anschlüsse im flachen Dach. In: Deutsches Dachdeckerhandwerk (DDH), Heft 14/69, S. 870–871.

Kakrow, K. H.: Detailausbildung in der Flachdachabdichtung. In: Deutsches Dachdeckerhandwerk (DDH), Heft 1/68, S. 25–30.

Lorenzen, Heinz: Auswertung der Erfahrungen aus der Sturmkatastrophe. In: Deutsches Dachdeckerhandwerk (DDH), Heft 4/69, S. 194–197.

Österreichisches Institut für Bauforschung: Abdichtungen und Abläufe bei Flachdächern, Dachterrassen, Balkonen, Loggien, Naßräumen; Forschungsbericht 39, 2. Auflage, Wien 1973.

Pfefferkorn, Werner: Konstruktive Planungsgrundsätze für Dachdecken und ihre Unterkonstruktionen. In: Das Baugewerbe, Heft 18/73,

S. 57–65; Heft 19/73, S. 54–59; Heft 20/73, S. 86–90; Heft 21/73, S. 54–63.

Praktisch, Theo: Flachdachanschlüsse mit Bitumendachbahnen. In: Deutsche Bauzeitschrift (DBZ), Heft 4/71, S. 689–690.

Rick, Anton W.: Sturmschäden in den USA. In: Das Dachdeckerhandwerk (DDH), Heft 9/70, S. 557–562.

Schaupp, Wilhelm: Das bituminöse Dach ist stets so gut oder so schlecht wie seine Anschlüsse. In: Deutsches Dachdeckerhandwerk (DDH), Heft 19/69, S. 1267–1282.

Schlenker, Herbert: Die Fachkunde der Bauklempnerei, A. W. Geutner Verlag, Stuttgart 1971.

Bearing surface and expansion joints

Balkowski, F. D.: Die Rißbildung am Deckenauflager. In: Das Dachdeckerhandwerk (DDH), Heft 2/74, S. 88–91.

Brandes, K.: Dächer mit massiven Deckenkonstruktionen – Ursache für das Auftreten von Schäden und deren Verhinderung. In: Berichte aus der Bauforschung, Heft 87, Verlag Wilhelm Ernst & Sohn, Berlin 1973.

Buch, Werner: Das Flachdach, Dissertation, Darmstadt 1961.

Grunau, Edvard B.: Probleme des Flachdachs. In: Deutsche Bauzeitung (db), Heft 11/73, S. 1262–1266.

Haage, K.; Kramer, C.: Neue Erkenntnisse über die Windbelastung auf Flachdächern. In: Das Dachdeckerhandwerk (DDH), Heft 22/74, S. 1446–1448.

Jaenke, H.: Sturmschäden an Dächern. In: Das Dachdeckerhandwerk (DDH), Heft 5/70, S. 264–279.

Kanis, H.: Das Flachdach im Sturm. In: Das Dachdeckerhandwerk (DDH), Heft 17/74, S. 1130–1132.

Kakrow, Helmut: Das flache Dach – Planung und Ausführung am Beispiel der Praxis. In: Deutsches Dachdeckerhandwerk (DDH), Heft 4/66, S. 144–147; Heft 9/66, S. 436–440.

Kramer-Doblander, Herbert: Temperaturspannungen in Flachdachkonstruktionen. In: Bitumen, Teere, Asphalte, Peche..., Heft 3/71, 21. Jahrgang, S. 93–96.

Paschen, H.: Bericht über die Untersuchung von Sturmschäden, die durch einen Tornado am 10. Juli 1968 im Raum Pforzheim hervorgerufen wurden. In: Das Dachdeckerhandwerk (DDH), Heft 4/73, S. 205–215.

Pfefferkorn, Werner: Konstruktive Planungsgrundsätze für Dachdecken und ihre Unterkonstruktionen. In: Das Baugewerbe, Heft 18/73, S. 57–65; Heft 19/73, S. 54–59; Heft 20/73, S. 86–90; Heft 21/73, S. 54–63.

Planckh, Rudolf: Die erforderliche Gleitfähigkeit der Flachdachschichten. In: Bitumen, Teere, Asphalte, Peche..., Heft 10/72, S. 408–410.

Rick, Anton W.: Windsog und Abreißen einer Dachhaut. In: Das Dachdeckerhandwerk (DDH), Heft 2/74, S. 106–112.

Rybicki, R.: Schäden und Mängel an Baukonstruktionen, Werner Verlag, Düsseldorf 1974.

Zellerer, Ernst; Thiel, Hanns: Probleme des Flachdaches beim statischen Entwurf. In: Die Bautechnik, Heft 2/70, S. 57–61.

Drainage

Buch, Werner: Das Flachdach, Dissertation, Darmstadt 1961.

Österreichisches Institut für Bauforschung: Abdichtungen und Abläufe bei Flachdächern, Dachterrassen, Balkonen, Loggien, Naßräumen; 2. Auflage, Wien 1973.

Soyeaux, H. J.: Rinnenabdichtung mit Polyisobutylen-Dachbahnen. In: Deutsches Dachdeckerhandwerk (DDH), Heft 18/69, S. 1158–1164.

Voorgang, H. J.: Die Entwässerung flacher Dächer I–IV. In: Deutsches Dachdeckerhandwerk (DDH), Heft 1/69, S. 14–16; Heft 3/69, S. 130–132; Heft 6/69, S. 278–282; Heft 9/69, S. 532–535.

Problem: Sequence of layers and individual layers

Flat roofs are used as roof terraces. This form of building construction was first used on a large scale in the residential buildings of the 1960s. Many architects, building contractors and manufacturers have limited experience in the correct structural formation of roof terraces.

Because of the working loads which have to be incorporated in the slab calculation, roof terraces are not normally constructed as lightweight single or double slab roofs, but rather as single slab structures on a reinforced concrete roofing slab.

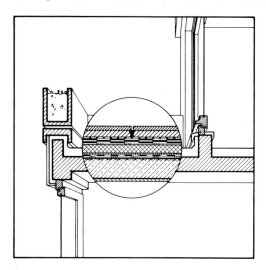

Analysis of failure in the typical cross-section of roof terraces shows a definite problem area in the surfacing and sealing layers. These defects can be traced back to the large stresses and to incorrect evaluation of the sealing function of the surface finish.

In many cases, it has been found that roof terraces and balconies do not fulfil the function of the single slab roof with respect to thermal technology and damp diffusion.

Common types of failure in the typical cross-section of roof terraces are illustrated on the following pages, with recommendations for their avoidance.

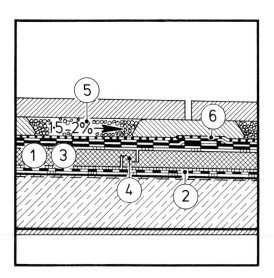

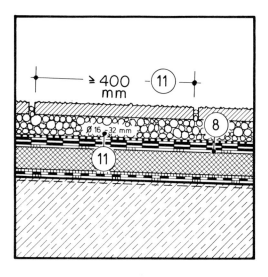

1 Roof terraces should have thermal insulation layers with a resistance to heat of $\geqslant 1\cdot3$ m² K/W, placed above the load-bearing roof slab. The insertion of additional thermal insulation layers under this is to be avoided (see C 1.1.2).

2 Under the thermal insulation there should be a vapour barrier with a vapour seal value (diffusion resistance equivalent air layer thickness) of $\geqslant 100$ m (see C 1.1.2).

3 It is essential that the compression strength of thermal insulation materials used in residential terraces be $\geqslant 200$ kN/m². In examples of large slabs with point supports or other heavy individual loads (plant containers) the minimum compression strength should be calculated in isolation or the load should be distributed over the largest possible bearing surface (see C 1.1.3, 1.1.5, 1.1.6).

4 The thermal insulation layer should be laid in two layers with joints lapped or folded over at their junction (see C 1.1.3).

5 Beneath the surfacing layer there must be a sealing layer. This must have a fall of $1\cdot5$–2% to the outlets. With steeper falls there is a danger of the surfacing material (e.g. flagstones) slipping on the separating or sealing layer. If this occurs, the surfacing layer must be secured (built-in cramps) (see C 1.1.4).

6 With falls of $1\cdot5$%, care should be taken to make certain that the finished sealing surface is even, e.g. that in the area of the adhesion flange of outlets there is no extra thickness which could prevent the water from flowing away (see C 1.1.4).

7 Roofing panels with linings which are unstable in damp conditions (crude felt and jute fabric) should not be used (see C 1.1.4).

8 Between sealing panels and the base there should be a stabilising layer which prevents a complete frictional connection. Where base slabs are large, the sealing layer should be fixed by spot adhesion and, where carried over the expansion joints, should remain unbonded for a width of 300 mm (see C 1.1.4).

9 In order to protect against possible mechanical damage of the sealing layer during subsequent work (scaffolding poles, falling of sharp-edged objects, movement of gravel, foot traffic, etc.), the finished surface should be applied directly after the completion of the waterproofing contract. If this is not possible, the sealing layer must be protected by timber planks or boards (see C 1.1.5).

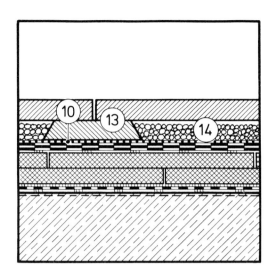

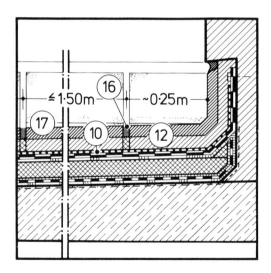

10 A frictional connection between the sealing layer and the surfacing layer must be prevented by the insertion of a durable, non-rotting separation layer. This can be achieved with two layers of loose-laid polyethylene foil (see C 1.1.5–7).

11 As the upper surface protective layer cannot be made permanently watertight, it should have porous joints and be laid on a layer which is also porous. For example, wearing surface layers made of natural or concrete flags of larger dimensions than 400 × 400 mm should be laid on a precompressed gravel bed ⩾ 50 mm thick consisting of washed granules of 16–32 mm dia. With greater loads these slabs should be increased in dimension and reinforced (see C 1.1.7 and 1.1.8).

12 Wearing surfaces consisting of flagstones or slabs on a mortar bed should be laid on 50 mm thick concrete with porous joints. With increased loading the individual weathering panels formed by expansion joints can be reinforced with structural steel mats (see C 1.1.7 and 1.1.8).

13 Weathering surfaces consisting of prefabricated concrete slabs or natural quarry stone can be laid on pads. In order to avoid excessive spot loads on the roof membrane the pads should be formed with as large a surface area as possible, i.e. in the form of mortar dabs (see C 1.1.7).

14 Dirt deposits in the hollow spaces under the raised surface may obstruct the passage of water over the sealing layer, and should be avoided by keeping the joint widths of individual slabs to a minimum, by frequent cleaning or by filling in the hollow spaces, e.g. with loose gravel (see C 1.1.5).

15 Asphalt coverings should be laid only on roof terraces which are protected from direct sunlight and those which are not subjected to spot loading (see C 1.1.8).

16 The weathering surface layer and its bedding must be separated by an edge joint from all adjacent structural members or structural members which pierce the surface layer, and must be subdivided by expansion joints into areas with a maximum length of approximately 1·5 m (see C 1.1.6).

17 The upper surface of the roof or terrace should be laid to a fall of 1·5–2% and connected to the drainage system (see C 1.1.8).

Roof terraces
Sequence of layers and individual layers

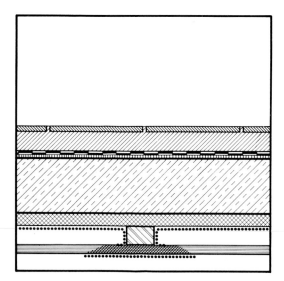

Where roof terraces or balconies are constructed without, or with only very thin, thermal insulation in their cross-section, water penetrates to the inner ceiling surface.

Where the internal thermal insulation is designed without the provision of a vapour barrier, the result, especially in damp rooms, is severe water penetration. The bonded thermal insulation layers separate from the adhesion surface, and the suspended ceilings warp.

Points for consideration

— Roof terraces should, in their dimensioning and the positioning of the thermal insulation and vapour barrier, be constructed as single slab flat roofs, for they are exposed to similar thermal and damp diffusion stresses.

— Inadequately dimensioned thermal insulation layers lead to high energy loss, strong temperature stress of the structural slab and such low surface temperatures that surface condensation can occur.

— The provision of thermal insulation without an internal vapour barrier within the roof section allows the water vapour to penetrate from the interior into the cold areas of the structural slab. The ensuing condensation can, especially in damp rooms, become so intense that the absorption capacity of the cross-section is exceeded and condensation appears on the under-side of the ceiling.

Recommendations for the avoidance of defects

● Roof terraces should have thermal insulation layers with a heat resistance of $\geqslant 1\cdot3$ m² K/W, placed above the loadbearing roofing slabs. The application of additional internal thermal insulation layers should be avoided.

● A vapourproof layer (equivalent air layer thickness > 100 m) should be placed under the thermal insulation layer.

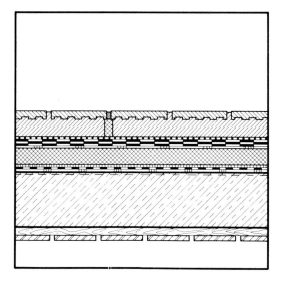

Roof terraces
Sequence of layers and individual layers

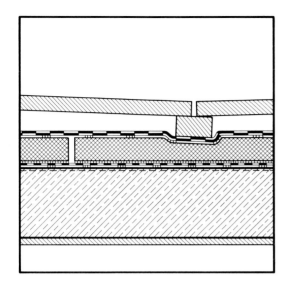

Flexible insulation layers under the terrace surface have been found to give way under loading. As a result of this, the surface slabs break and become uneven, especially where prefabricated slabs have been laid loose on pad bases, with resulting damage to the sealing layer under the pad bases.

If the thermal insulation layer is not continuous over the roof terrace, patches of surface saturation can be seen on the under-side of the ceiling.

Points for consideration

- The thermal insulation layer on top of a roof terrace may – especially where large roofing slabs on pad bases are used – be exposed to high compressive loads. These loads must be carried with little or no movement and without failure of the structure.

- With single-layer, butt-jointed insulation slabs, defective areas forming thermal bridges can easily occur if construction is inaccurate or damaged slabs are used.

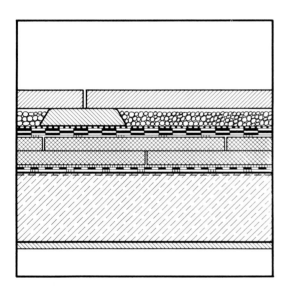

Recommendations for the avoidance of defects

● The thermal insulation materials selected for roof terraces should be such that the applied loads can be absorbed without damage. Large slabs on pad bases or other point loads (plant containers) must be taken into account in the load. The minimum compression strength of the thermal insulation layers in terraces of residential buildings should be about 200 kN/m².

● Thermal insulation layers should be laid in two layers with joints lapped or folded over.

Roof terraces
Sequence of layers and individual layers

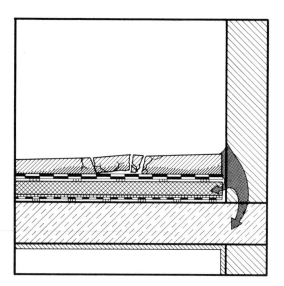

Under the weathering surface layer, sealing layers which have been assumed to be watertight are frequently found to be defective. Sometimes sealing layers are not carried right to the edge of the terrace, or they have an inadequate fall, resulting in damage to the roof structure or damage to the surface layer caused by water penetration and freezing (see C 1.1.6 – Stressing of surfacing).

Moreover the sealing layer itself may show, in addition to damage from mechanical stresses (see C 1.1.5 – Stressing and protection of sealing layer), the following defects:
- Signs of rotting in sealing panels which are unstable in wet conditions;
- Seepage at the overlap joints;
- Cracks in sealing panels wholly bonded to the base, especially where large glass-fibre slabs which are inherently unstable are used as a base.

Points for consideration

- Surface layers, even flagstone or mastic surfacings, cannot be made permanently waterproof. Therefore under the surface finish of roof terraces, as on non-trafficked roofs, a sealing layer must be provided which has the function of protecting the structure from penetration by surface water and, in addition, of ducting away water that penetrates the surface finish and diverting it into the drainage outlet.
- It is not easy to ensure the waterproof properties of roof terraces with heavy and fixed upper surface slabs, or surfacing materials connected to the sealing layer. It is necessary to take great care in the physical operation of laying the sealing layer, and also in the selection of suitable materials.
- Standing water on the sealing layer increases the danger of penetration into the structure below, especially where the bonding of the overlap joints is defective.
- Standing water on the sealing layer may lead to failure of the surface finish as a result of efflorescence or the freezing of saturated slabs.
- Increased thickness caused by overbonding of the sealing layer with the flanges of roof outlets may prevent the free drainage of water.
- Sealing layers wholly bonded to the base are completely subjected to the movements of the base and may be over-stressed.
- The sealing layer is particularly strongly stressed over the expansion joints of large base slabs, where greatest movements occur.
- The trapping of damp and air under the roof sealing membrane, caused by a wet and dirty base, leads in intense heat (e.g. under a slab surface which rests on pad bases) to excess pressure and, in the case of defective adhesion, to the formation of bubbles.

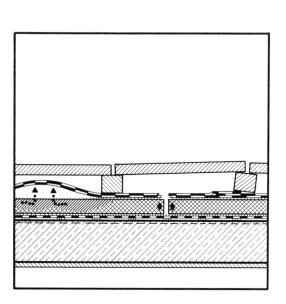

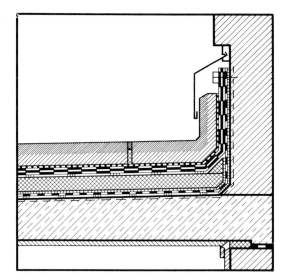

Recommendations for the avoidance of defects

● The sealing layer must have a fall of 1·5–2% to the outlets. With larger falls there is danger of the surfacing (e.g. flagstones) slipping on the separating layer. If necessary an anchorage (abutment, embedded angle, etc.) for the upper surface slabs should be provided.

● The recommendations laid down for single slab roofs are applicable to the laying of the sealing layer (see A 1.1.9 – Laying of sealing layer).

● Where a roof terrace has a minimal fall, care should be taken that the finished sealing surface has no excessive variation in level (caused by the overbonding of the roof membrane with the flanges or connection foils of the outlets) which might prevent the water from flowing away.

● Roofing panels with linings which are not stable in damp conditions (crude felt and jute fabrics) should not be used.

● There should be a stabilising layer between the sealing panels and the base to prevent a complete frictional connection. In examples of large base slabs in addition to a spot adhesion, the sealing layer should remain unbonded for a width of 300 mm over the expansion joints. This can be achieved by the insertion of a loose strip (see A 1.1.7 – Stabilising layer – Adhesion of roof membrane to base).

Roof terraces
Sequence of layers and individual layers

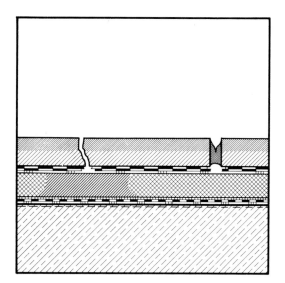

In roof terraces the sealing layer may suffer damage from various causes, resulting in the penetration of rainwater to the structure below.

Direct mechanical damage may be caused by workmen (especially when laying flagstones or slabs) working on the unprotected sealing layer. Frequently, however, pressure points and cracks are due to the lack of a separating layer or a load-distributing layer between the sealing layer and the upper surface (flagstones, slabs, or slabs on pad mounting).

Points for consideration

— Sealing layers, whether bituminous panels or synthetic foils, are vulnerable to mechanical damage, especially when the sealing base consists of soft flexible thermal insulating material or where there are voids under the sealing layer caused by open expansion joints or unevennesses in the base.

— The sealing layer is particularly subject to damage if frictionally connected with an upper surface weathering finish which has considerable thermal movement.

— Excessive point loads, e.g. heavy slabs on pad supports, cause pressure points on the sealing layer, if a bituminous waterproof layer with a low fusion point is used and high temperatures build up under the upper surfacing layer (e.g. on a south-facing terrace).

— Rigid pad supports made of synthetic material or metal cannot adequately compensate for the unevennesses in the sealing layer caused by overbonding.

— Between the pad-supported weathering surface and the sealing layer, dirt and leaf deposits accumulate from the surrounding area and these can lead to intense ponding. The free drainage of water is thus prevented. In addition, these voids provide space for vermin.

Recommendations for the avoidance of defects

● In order to protect against possible mechanical damage resulting from subsequent works, the upper surface should be applied directly after completion of the sealing layer. If this is not possible, the sealing layer should be protected by means of a working platform formed of timber planks or boards.

● Where flagstones or precast concrete slabs are used, a separating layer must be provided to prevent the transfer of movement between the sealing layer and the weathering surface.

● Excessive point loading, originating, for example, from pad-supported slabs, must be avoided. Wide surface bearings, such as mortar beds, distribute the load uniformly and thus reduce the risk of damage.

● Individual slabs on pad supports should, in order to reduce dirt deposits in the voids below, be laid with minimal joint widths (see also C 2.3.3) and/or the voids should be filled with loose gravel.

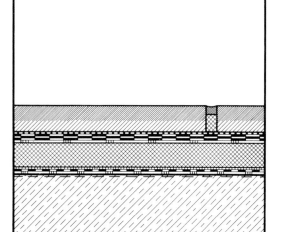

Roof terraces
Sequence of layers and individual layers

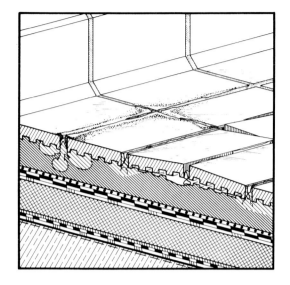

The weathering surface layers of roof terraces reveal cracks in the surface, both in the general area and also at restrained points (e.g. protruding wall corners), if they are laid continuously without subdivision by expansion joints. These cracks rapidly reveal signs of damage to the mortar, and the build-up of dirt causes further problems. Damage is also caused by the saturation of the roof structure (thermal insulation layer) or of the interior, as in the region of the cracks in the surface layer the sealing layer is overstressed, and fractures.

The surface layer sometimes works loose in places, and bulges. There also occurs failure of the edge joints and, in asphalt surfaces, edge fractures. In the angle formed with the vertical surface of a structural member (parapet, wall) where the flagstones are applied abutting the edge and where pad-supported slabs are laid close together, deformation and failure of the sealing layer and extensive dampness are found (see C 2.1.5 – Sealing junctions made of foils or strips).

Points for consideration

— Because they are exposed horizontally on the building, with the thermal insulation layers normally laid directly below, the weathering surface layers are exposed to considerable variations of temperature. Additionally, there are varying dampness stresses.

— The resultant changes in form and length, when restricted by adjacent components or those piercing the surface, lead to tensile stresses which, because of the minimal strength and material properties (above all, the low tensile strength) of the surface layer, cause cracks and fractures.

— Inserted reinforcement (light structural steel mesh) improves the ability to absorb loads (bending moments).

— The considerable linear expansion caused by changes in temperature of asphalt surfacings, when restricted by vertical obstructions such as abutting walls, develops a strong compression strain in the asphalt (plastic deformation). With falling temperatures (reducing plasticity) contractions occur which result in settlement cracks in adjacent structural members (walls) or, in continuous surfacing having a defective separation from the base (sealing layer), cracks appear in the surface.

— As the linear expansion of the weathering surface layers is unavoidable, transfer of the resultant force to the sensitive sealing layer must be prevented.

— Surface finishes laid directly on to the sealing layers form, in time and with temperature influence, an internal bonding, e.g. with the bituminous adhesive layers of the sealing layer.

Recommendations for the avoidance of defects

● A frictional connection between the sealing layer and surface finish must be prevented. To achieve this, a permanently effective separation layer should be inserted, in the form of two layers of loose-laid polyethylene foil.

● The surface finish must be separated from all adjacent or penetrating structural members by an edge joint and be subdivided by expansion joints in surface areas longer than 1·5 m (see C 2.3.3 – Expansion joints in the structure).

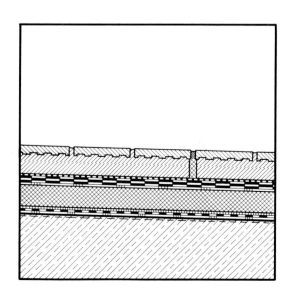

Roof terraces
Sequence of layers and individual layers

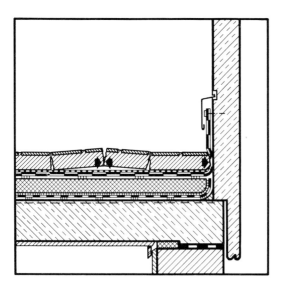

On roof terraces, complete saturation of the roof structure (thermal insulation layer) and the interior is sometimes found, because the structural parts have been designed with no sealing layer above the thermal insulation. The 'watertight' surface layers of asphalt and cement flagstones, sometimes given a waterproof coating film, may leak because they are stepped on after settlement of the edge, or because cracks and fractures develop in the surfacing, coating and expansion joints.

Where the sealing layer has been laid level or with only a slight fall, or is not connected to the drainage outlets, damage to the surface material results. Rainwater penetrates into the surfacing through cracks and expansion joints. If 'waterproof' weathering layers are of natural stone, ceramic slabs, etc., the surface may remain saturated for a long time. In the area of cracks and joint edges, mortar is eroded and dirt deposited. Similarly, the joints loosen and fail; where the weathering layer is of impervious slabs, these may become dislodged. These surface defects increase with traffic on the roof terrace surface.

Points for consideration

– Surface finish layers of roof terraces cannot be made watertight.
– Dampness which has penetrated through a 'waterproof' surface as a result of localised leaks in the weathering layers is slow to dry out in the area of penetration. Mortar particles are bound in solution with dirt deposited on the upper surface. In frosty weather the entrapped moisture results in freezing of the upper surface layer and leads to joint failures.
– The aim must be to remove, as soon as possible, any rainwater which penetrates the surfacing material.

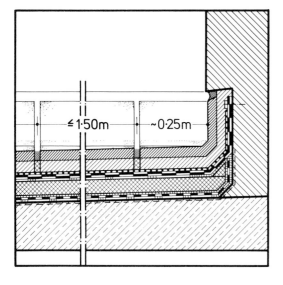

Recommendations for the avoidance of defects

● The surfacing material should be applied on a sealing layer laid to fall with an intermediate separating layer.
● The upper surface layer (weathering layer) should be applied on layers which are permeable to water.
● Weathering layers made of flagstones or slabs laid in a mortar bed should be set on 50 mm thick protective concrete with porous joints. For greater loads, the individual surfacing areas formed by expansion joints can be reinforced with structural steel mesh.
● Weathering surfaces consisting of precast concrete slabs or natural stone larger than 400 × 400 mm should be set into a precompressed gravel fill ⩾ 50 mm thick, composed of washed granules of 16–32 mm dia. For larger loads, these slabs should be sized and reinforced accordingly.
● Weathering surfaces consisting of precast concrete slabs or natural stone can be laid on pads. In order to avoid excessive point loads on the roof membrane, the pad bases should be made with as large a surface area as possible, e.g. in the form of mortar beds.

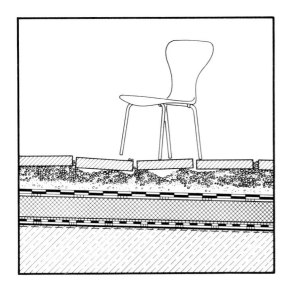

Use of roof terraces is found to be seriously impeded by various defects of the weathering surface. In the majority of terraces, the roof surface has little or no fall, so that rainwater and melting snow lead to damp penetration, formation of puddles and icing. Where the asphalt surface is exposed to sunlight, the surface softens and, as a result, pressure points develop which penetrate the whole thickness of the surfacing material (e.g. under tables and chairs). In applied coatings, cracks, bubbles and separations develop; flagstone and slab surfaces crack; along these cracks and along the edges of expansion joints are found erosion and dirt deposits. At the edges of structural members the surface layer loosens, e.g. at the plate connection. Following periods of frost, slab surfaces are found to be only partly supported, with voids beneath them. Where precast slabs are set loosely in a sand or gravel bed, they and the joints become uneven.

Points for consideration

— If the fall of the upper surface of the roof is not sufficient or if the catchment area of the roof outlets is too large, the rainwater drains away slowly or remains as puddles until it evaporates. In frosty weather it can form ice on the upper surface. The aim must be to drain rainwater away as fast as possible from the upper surface.

— Precast slabs set loose in a gravel bed can tilt on the flexible base; equally, the gravel bed can undergo erosion of small granules and undermining, resulting in voids under individual slabs.

— The surfacing, especially the weathering layer, should be constructed with a view to its specific use. For example, where furniture is to be placed on the terrace the fall must be ≤ 2%. For increased loads, e.g. plant containers, furniture and outdoor equipment, an appropriate loading capacity and stability of the surfacing must be provided. Similarly, the compression strength of the sealing base (thermal insulation layer) must be considered (see C 1.1.3 – Compression strength and treatment of thermal insulation).

— Asphalt surfaces have a tendency, under long-term point loads and especially at higher temperatures (sunlight, concentration of heat), to deflection and lateral movement at pressure points. This can only be restricted by the selection of appropriate materials.

Recommendations for the avoidance of defects

● The upper surface of the weathering surface must be laid to a fall of 1·5–2% and be connected to the drainage. Where slabs are laid in a gravel bed or on pads, the fall and drainage connection to the roofing surfaces are not necessary.

● The surfacing should be dimensioned according to the applied loads and reinforced if necessary. Reinforcement (structural steel mesh bedded) in protective concrete or a mortar bed for slab surfacing must be laid separated from the individual surfacing zones formed by the expansion joints.

● Loose-laid precast slabs should be as large as possible (≥ 400 × 400 mm) and heavy, and be set on a precompressed gravel layer ≥ 50 mm thick made of washed granules of 16–32 mm dia, so that their whole surface is supported.

● Asphalt surfaces should be laid only on roof terraces which are protected from direct sunlight and are not subject to great stress from point loads.

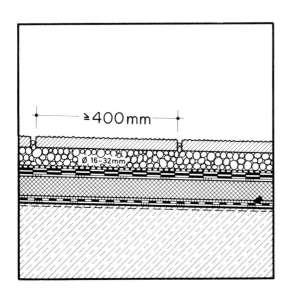

Roof terraces
Typical cross-section

General texts and principles

Eichler, Friedrich: Bauphysikalische Entwurfslehre, Band 2, 4. Auflage, Verlagsgesellschaft Rudolf Müller, Köln 1973.

Hanson, Rune: Takterrasser – Tätskikt och Skyddsbeläggning, S NB Rapport 62, Stockholm 1960.

Lufsky, Karl: Bauwerksabdichtungen – Bitumen und Kunststoffe in der Abdichtungstechnik, 2. Auflage, B. G. Teubner, Stuttgart 1970.

Moritz, Karl: Flachdachhandbuch – Flache und flachgeneigte Dächer, 4. Auflage, Bauverlag Wiesbaden und Berlin 1975.

AGI-Arbeitsblätter A 10 – Industrieböden, Hartbetonbeläge; A 11 – Industrieböden, Zementestrich als Nutzboden.

DIN 4108 – Wärmeschutz im Hochbau, Ausgabe 1969 mit den ergänzenden Bestimmungen, Oktober 1974.

DIN 4109 – Schallschutz im Hochbau, Blatt 4 – Schwimmende Estriche auf Massivdecken, September 1962.

DIN 4122 – Abdichtung von Bauwerken gegen nicht drückendes Oberflächenwasser und Sickerwasser mit bituminösen Stoffen, Metallbändern und Kunststofffolien, Richtlinien, Juli 1968.

DIN 18337 – VOB/C Abdichtung gegen nicht drückendes Wasser, Februar 1961.

DIN 18338 – VOB/C Dachdeckungs- und Dachabdichtungsarbeiten, August 1974.

DIN 18353 – VOB/C Estricharbeiten, August 1974.

DIN 18354 – VOB/C Asphaltbelagarbeiten, Februar 1961.

Thermal insulation and vapour barrier layers

Buch, Werner: Grundlegende bauphysikalische Fragen des Flachdaches unter besonderer Berücksichtigung des Wasserdampfproblems. In: Bitumen, Teere, Asphalte, Peche..., Heft 10/71, S. 393–401.

Caemmerer, W.: Berechnung der Wasserdampfdurchlässigkeit und Bemessung des Feuchtigkeitsschutzes von Bauteilen. In: Berichte aus der Bauforschung, Heft 51, Verlag Ernst & Sohn, Berlin 1968.

Cammerer, W. F.: Der Wärme- und Diffusionsschutz des Flachdaches nach dem heutigen Forschungsstand. In: Bitumen, Teere, Asphalte, Peche..., Heft 10/71, S. 402–412.

Götze, Heinz: Dämmschichten für Dächer und Außenwände. In: Das Bauzentrum, Heft 3/72, S. 65–85.

Haushofer, Bert; Wichmann, H.: Vollwärmeschutz aus der Sicht des Dachdeckers. In: Das Dachdeckerhandwerk (DDH), Heft 19/74, S. 1256–1262.

Rick, Anton: Grenzen der Wärmedämmung von Flachdächern. In: Bitumen, Teere, Asphalte, Peche..., Heft 25/74, S. 360–363.

Rick, Anton: Sind Dampfsperren wirklich unnötig? In: Deutsches Dachdeckerhandwerk (DDH), Heft 14/69, S. 872–873.

Riedel, Peter: Schaumglasdämmungen ohne Dampfsperre. In: Deutsches Dachdeckerhandwerk (DDH), Heft 2/67, S. 83–85.

Sealing layer: stress, protection, laying

Baumgärtner, Konrad: Begehbare Dachterrassen und deren Abdichtung. In: Bitumen, Teere, Asphalte, Peche... (BTAP), Heft 3970, S. 112–118.

Gäbges, Peter: Die Fragwürdigkeit von Plattenkunststofflagern auf abgedichteten Terrassen. In: Deutsches Architektenblatt (DAB), Heft 11/71, S. 404–405.

Haefner, Rudolf: Die Abdichtung von unterkellerten Hofdecken, Terrassen über Nutzräumen und Flachdächern. In: Boden, Wand + Decke, Heft 6/66, S. 482–500.

Osterritter, Kurt: Platten, Estriche, Abdeckungen über bituminösen Trägerabdichtungen. In: Bitumen, Teere, Asphalte, Peche..., Heft 10/65, S. 448–450.

Schild, Erich: Verhütung von Konstruktions- und Ausführungsschäden an Flachdächern. In: Bitumen, Teere, Asphalte, Peche..., Heft 10/68, S. 386–389.

Schütze, Wilhelm: Der Estrich auf Dächern und Terrassen. In: Boden, Wand + Decke, Heft 3/65, S. 182–192; Heft 4/65, S. 294–304.

Zentralverband des Dachdeckerhandwerks: Richtlinien für die Ausführung von Flachdächern, Ausgabe Januar 1973, Helmit Gros Fachverlag, Berlin 1973.

Surface finish of roof terraces – Design, nature, linear expansion

Gäbges, Peter: Die Fragwürdigkeit von Plattenkunststofflagern auf abgedichteten Terrassen. In: Deutsches Architektenblatt (DAB), Heft 11/71, S. 404–405.

Haefner, Rudolf: Die Abdichtung von unterkellerten Hofdecken, Terrassen über Nutzräumen und Flachdächern. In: Boden, Wand + Decke, Heft 6/66, S. 482–500; Heft 7/66, S. 588–596; Heft 8/66, S. 684–696.

Hoch, Eberhard: Flachdächer mit harter Schale – Terrassendächer und Parkdachabdichtungen. In: Das Dachdeckerhandwerk (DDH), Heft 11/72, S. 788–792.

Lufsky, Karl; Konzack, Kurt: Begehbare Dachterrassen. In: Bitumen, Teere, Asphalte, Peche..., Heft 10/67, S. 359–368.

Moritz, Karl: Estriche auf Flachdächern. In: Bitumen, Teere, Asphalte, Peche..., Heft 3/65, S. 116–121.

N. N.: Fliesen als Balkon- und Terrassenbeläge. In: Deutsches Architektenblatt (DAB), Heft 14/73, S. 1145–1146.

Rick, Anton W.: Einiges über Asphaltbeläge auf Terrassen. In: Bitumen, Teere, Asphalte, Peche..., Heft 2/72, S. 82–83.

Rick, Anton W.: Risse und Randabsetzungen bei Asphaltterrassenbelägen. In: Bitumen, Teere, Asphalte, Peche..., Heft 10/73, S. 422–423.

Schütze, Wilhelm: Der Estrich auf Dächern und Terrassen, eine Estrichart, die ganz besondere Sorgfalt erfordert. In: Boden, Wand + Decke, Heft 3/65, S. 182–192; Heft 4/65, S. 294–304.

Schulze, Karl: Technologische Besonderheiten der Gußasphalt-Estriche. In: Boden, Wand + Decke, Heft 9/65, S. 733–742; Heft 10/65, S. 817–821.

Zimmermann, Günther: Dachterrassen mit aufgestelzten Belägen – Durchfeuchtungen von Decken und Wänden. In: Deutsches Architektenblatt (DAB), Heft 12/74, S. 927.

Zimmermann, Günter: Plattenbeläge auf Balkonen. In: Architekt + Ingenieur, Heft 5/71, S. B1–B4.

Problem: Connection to vertical abutments

With roof terraces, continuity of construction of the structural members with that of the adjoining vertical structural wall or beam must always be ensured: for example, in a staggered form of construction (terraced houses), connection to the external walls of the adjoining living units; in other cases, junction with built-in balustrading, plant containers, etc.

The primary work of the junction with the vertical structural member is to ensure the continuity of the sealing layer at the angle with the roof terrace itself, and to prevent the overstressing of the vertical wall surface through dampness.

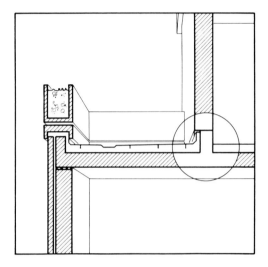

The findings of the research project into building damage show these connections to walls or abutments to represent the most serious problem in roof terrace design: of all instances of structural damage, one in four was the result of faulty connection to vertical surfaces.

Analysis shows that this high incidence of defects originates in a few, frequently recurring errors in the design or construction of roof terraces.

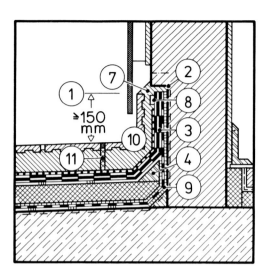

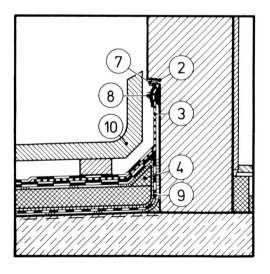

1 The sealing layer, at its connection to the vertical structural surface, must be raised above the highest water level that may be expected in adverse conditions. Thus the sealing surface must extend ⩾ 150 mm above the upper edge of the terrace surface layer (see C 2.1.2).

2 The required structural height of the sealing edge must be established at design stage. This is calculated from the necessary upstand height and the number and thickness of individual layers – surface finish, roof structure, falls and underlays (see C 2.1.2).

3 The angled slope and the junction of the sealing layer to the vertical wall should be built up from the roof membrane itself or by the use of connection foils of similar properties and behaviour (see C 2.1.4).

4 The transition of the sealing layer to the angled slope should be as uniform as possible, avoiding abrupt bends at change of plane; the gradual transition must be preformed in the sealing base by the use of triangular wedges made of thermal insulating material, ensuring a continuous sealing base (see C 2.1.5).

5 The formation of the sealing angle by using wall connection plates should be avoided (see C 2.1.4).

6 If the use of connection plates is unavoidable, they should have a protective covering (e.g. bitumen) and should be provided with sliding seams, to allow for expansion, at adequate centres (with titanium sheet zinc, at least every 3 m). After the removal of dampness and dirt, the plate should be bonded in horizontally ⩾ 120 mm between the multiple layers of the sealing layer (see C 2.1.4).

7 Where the angled sealing layer joins the vertical surface, its end must be protected from water streaming down the wall. This can be achieved by recessing the end of the sealing layer behind the plane of the upper water-diverting surface, or by protecting it with a cladding element (see C 2.1.3).

8 In order to avoid sagging of the sealing angle, it should be fixed along its whole upper length and should be protected from direct sunlight (see C 2.1.3).

9 The roof structure above the loadbearing slab, especially the thermal insulation layer, must be additionally protected against damp penetration by firm adhesion of the vapour barrier (see C 2.1.3).

10 The angled pliable sealing layer must be protected from thermal movement and damage by the formation of a skirting with the surface slabs or by suspending a metal cover flashing in front of it (see C 2.1.4).

11 The surface slabs of the roof terrace must be separated from the angle with the wall by an edge joint, in order to prevent transmission of movement from the surface material to the angled sealing layer (see C 2.1.5).

Roof terraces
Connection to vertical abutments

By far the most frequent cause of serious damp failure of the roof structure (thermal insulation) and of the inner building below is the complete lack of, or inadequate upstand height of, the sealing layer, at its junction with the vertical surface, above the finished roof level. Often the angled sealing layer is found to end at roof surface level, and to be fixed only with mastic. In other roof terraces the intended height is reduced to only a few centimetres above walking surface, as in establishing the upstand height at the design stage the actual structural depth of the slab, the surface finish and falls were not taken into account.

Points for consideration

— The edge of the angled sealing layer at its junction to the vertical wall surface cannot be permanently sealed, in practice, against water penetration. Therefore it must be protected.

— In the region of the roof terrace surface, the abutting vertical surface is exposed to major stresses from the build-up of water and spray, comparable to those occurring at ground level. Therefore measures to protect against dampness are necessary.

— Limited or imperfect falls to the upper surface of the terrace augment the problem of water pressure and penetration of the surface covering.

Recommendations for the avoidance of defects

● At junctions to vertical wall surfaces the sealing layer must be raised above the highest water level that may be expected in the worst possible circumstances. Thus, the sealing layer must extend ≥ 150 mm above the finished surface level.

● The required structural height of the sealing edge must be established at design stage. This is calculated from the necessary upstand height and the number and thickness of individual layers — surface finish, roof structure, falls and underlays.

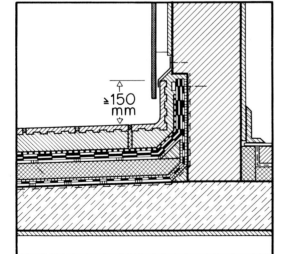

Roof terraces
Connection to vertical abutments

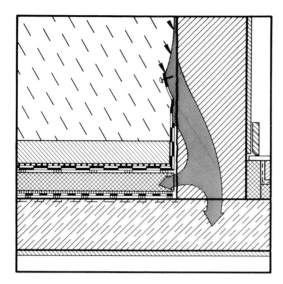

Rainwater may penetrate behind the edge of the raised sealing layer at the connection to the vertical wall surface, causing damage to soft sealing layers, which form a protective upstand to the external surface of the wall and are fixed only by adhesion. The adhesive strips come loose in places, sometimes over considerable lengths. However, frequently no further sealing protection is provided for the exposed edges.

Points for consideration

— The raised sealing layer is subject, at the junction to the slope, to strong thermal influences which reduce its stability and that of the adhesive layers.

— The higher a vertical surface and the more exposed it is to driving rain, the more intense is the stream of water which runs down this surface.

— A junction of the sealing layer designed to lie in front of the main vertical wall plane forms an obstacle to the flow and is particularly vulnerable to water penetration.

— The surface of the vertical wall is not completely even, so that a pressed metal flashing can find no uniform and continuous bearing surface. Sealing by cement has very limited durability.

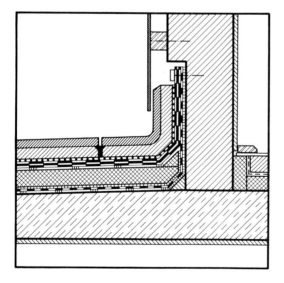

Recommendations for the avoidance of defects

● At the junction of the angled sealing layer to the vertical wall surface the end of the sealing must be protected from the increased velocity of the water flow, by recessing the end of the sealing layer behind the plane of the water-diverting upper surface, or by protecting it with a cladding element.

● In order to avoid sagging of the sealing angle, this should be continuously fixed along its whole upper length and should be protected against direct sunlight.

● The roof structure above the loadbearing slab, especially the thermal insulation layer, should be additionally protected on the lateral surface against damp penetration by the firm adhesion of the vapour barrier.

Roof terraces
Connection to vertical abutments

Where metal profiles are used to form the sealing angle at the junction with the vertical surface (wall, parapet), they frequently let in water. Long lengths of sheet metal without provision for expansion have been found to exhibit torn soldered seams, severe buckling and warping. In some instances the bonding of the profiles with the sealing layer is not waterproof because the metal has been bonded on only one side to the sealing layer. Where sections of lead or zinc are designed without a protective layer and laid in the mortar bed, ageing and corrosion of the metal causes damage and leaking.

Points for consideration

— Angles of the sealing layer which are not protected (e.g. by a coved slab) are exposed to the weather and thus to variations in temperature. Depending on the material, considerable fluctuations of length or form can be generated.

— The use of pressed metal angles requires a direct permanent bonding of two very different structural materials by means of an adhesive.

— The changes in length can be reduced by subdividing the metal angles into shorter lengths and providing expansion joints.

— Metals — especially sheet lead and zinc — are vulnerable chemically to fresh lime or cement mortar and concrete in that the calcium hydroxide dissolves the metal. Direct contact must be avoided.

Recommendations for the avoidance of defects

● The angle and junction of the sealing layer to the vertical structural wall should be formed by the roof membrane itself or by use of connection foils of similar material behaviour.

● The angled sealing layer must be protected from thermal movement and damage by raising the surface slabs or by suspending metal or other lightweight claddings in front of it.

● The formation of the sealing angle by means of sections directly fixed to the wall should be avoided.

● If the use of pressed metal angles is unavoidable, they should be protected by a surface coating (e.g. bitumen) and provided with expansion devices at adequate centres (with titanium sheet zinc, $\geqslant$ every 3 m) to allow for movement. After the removal of damp and dirt (oil, etc.) the metal profile should be bonded for a depth of $\geqslant 120$ mm between the individual layers of the multi-layer sealing membrane.

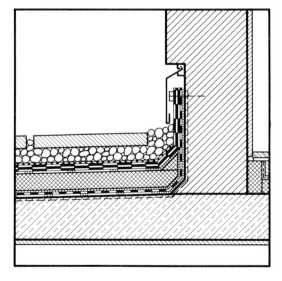

Roof terraces
Connection to vertical abutments

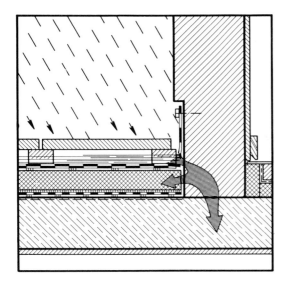

Where the angle is formed from the sealing layer itself or from other pliable foil (e.g. from synthetic material), failure often results from mechanical stress. At junctions where the sealing layer is diverted at a right angle and continuously unsupported through the change of planes, the material may be pierced. In other examples, surface slabs are laid tight against the angle (i.e. without separating expansion joints) and the sealing layer is damaged by slab movement.

Points for consideration

— In the course of the construction of the surface finish the unprotected sealing layer is liable to severe mechanical stress caused by foot traffic, wheelbarrows, storage of construction materials, etc.

— The vulnerability of the sealing layer to mechanical stress is considerable in those areas where it is unsupported (see C 1.1.5 – Stressing and protection of sealing layer).

— The surface layers are exposed to extreme variations in temperature and are therefore subject to severe thermal linear expansion. This linear expansion takes effect at the edge of the continuous roofing surface, i.e. at the junction to the vertical surfaces, where the vulnerable sealing layer is angled (see C 1.1.6 and 2.3.4 – Expansion joints in the roof terrace surface).

Recommendations for the avoidance of defects

● The transition of the sealing layer to the angled slope should be as uniform as possible, avoiding abrupt bends at the change of plane; the gradual transition must be preformed in the sealing base by the use of triangular wedges made of thermal insulating material, ensuring a continuous sealing base.

● The surface slabs of the roof terrace must be separated from the angle with the wall by an edge joint, in order effectively to prevent stressing of the angled sealing layer by the expansion of the surface material under thermal influences.

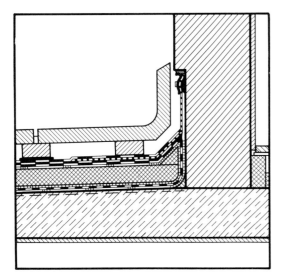

Problem: Edge of structural member

The structural design of the edges of structural members is dependent upon the method of drainage of the terrace roof surface. In examples of external drainage by means of gutters, an uninterrupted edge detail must be formed; where drainage is internal by means of individual outlets, an upstand edge is required.

The edge design is equally influenced by the required parapet or railing design, which may alternatively take the form of plant containers.

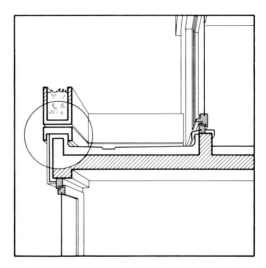

Damaged roof terraces are frequently found to have defects to the edges of structural members with internal drainage points. This is caused by a lack of fall or inadequate fall of the sealing layer at the edge of the structural member. Frequently damp penetrates to the sealing layer.

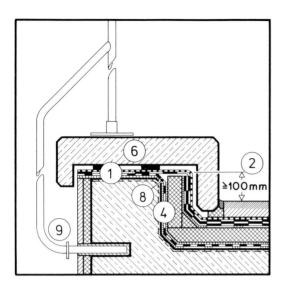

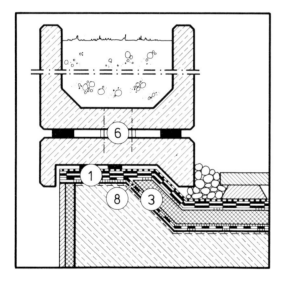

1 The sealing layer must be brought to the edge of the structural member and raised to the required height of at least 100 mm (see C 2.2.3).

2 At the edge of inwardly drained roof terraces, the sealing layer must be raised above the highest water level that may be expected, and to $\geqslant 100$ mm above the upper surface of the terrace roof (see C 2.2.2).

3 Where edges are sloped, the sealing layer itself should be raised on a slope preformed in the sealing base (see C 2.2.4).

4 The required structural height of the edge of the structural member must be established at design stage, and is calculated from the necessary upstand height and the number and thickness of individual layers – surface finish, roof structure, falls and underlays (see C 2.2.2).

5 The sealing layer may be connected to parapet units only when these are permanently sealed against rainwater. The sealing layer must be raised to a height of 150 mm and be secured with a cover flashing to the sealing edge (see C 2.2.3).

6 If the parapet units at the edge of the structural member (e.g. plant containers) are not permanently watertight or have joints which are difficult to seal, the sealing layer must go underneath those units and up to the edge of the structural member (see C 2.2.3).

7 A frictional connection of the sealing layer with metal edge profiles should be avoided. Metal sections should only be used as cover elements (see C 2.2.4).

8 The roof structure above the loadbearing slab (thermal insulation layer) should be further protected at the external surface against possible dampness, e.g. by continuing the vapour barrier around it (see C 2.2.2).

9 Balustrades should be secured outside the sealed surface in the loadbearing structure (e.g. exposed edge of the reinforced concrete roofing slabs) (see C 2.2.5).

10 Balustrade fixings on the weathering surface of the roof slab, which pierce the surface finish and sealing layer, should be avoided (see C 2.2.5).

11 If a fixing on the weathering surface of the roof terrace cannot be avoided, the balustrade can be secured in coping units – e.g. precast concrete units – which are laid on top of the sealing layer (see C 2.2.5).

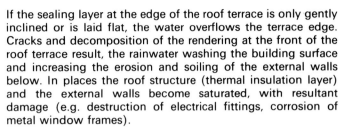

Roof terraces
Edge of structural member

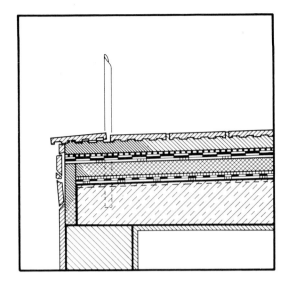

If the sealing layer at the edge of the roof terrace is only gently inclined or is laid flat, the water overflows the terrace edge. Cracks and decomposition of the rendering at the front of the roof terrace result, the rainwater washing the building surface and increasing the erosion and soiling of the external walls below. In places the roof structure (thermal insulation layer) and the external walls become saturated, with resultant damage (e.g. destruction of electrical fittings, corrosion of metal window frames).

Defective edge construction has a more marked effect in roof terraces where sealing layer and surface slabs have been laid without, or with a minimal, fall.

Points for consideration

— In roof terraces, according to the selected surface construction, rainwater can be drained either from the upper surface of the roof or from the sealing layer.

— Where the fall on the roof is minimal or defective, the formation of puddles or wind pressure can cause an increased flow of water at the edge of the structural member.

— Backwash or blocked outlets can increase the water pressure.

— If the edge of the roof structure (thermal insulation layer) is not provided with a sealing layer, then rainwater penetrating through joints or cracks can lead to large areas of dampness in the roof structure.

Recommendations for the avoidance of defects

● On inwardly drained roof terraces the edge of the sealing layer must be taken above the highest level of water to be expected in unfavourable conditions. This height must be ⩾ 100 mm above the roof surface.

● The required structural height of the edge of the roof slab must be established at design stage. This is calculated from the required height of the upstand and the number and thickness of individual layers – surface slab, roof structure, falls and underlays.

● The roof structure above the loadbearing slab, especially the thermal insulation layer, must be additionally protected at the vertical face against possible damp penetration by the complete adhesion of the vapour barrier.

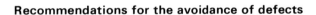

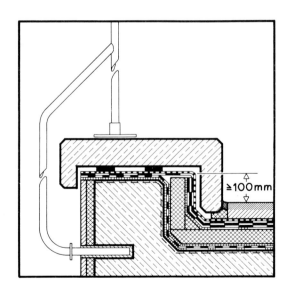

⩾100 mm

Roof terraces
Edge of structural member

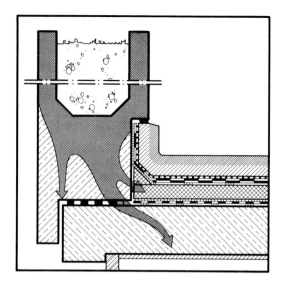

Continuously along the junction of the sealing layer with the parapet units damage may result from water penetrating the roof structure (thermal insulation) and to the rooms below. This is caused by inadequately protected and incorrectly fixed ends of the sealing layer, which come loose from the base – usually because they have been designed to have little or no height above the upper surface of the roof, so that water can run behind them.

If plant containers are placed at the edge of the roof terrace to which the sealing layer is fixed, damage by damp penetration has been found to be particularly extensive. The rainwater penetrates through the plant containers into the actual roofing slabs, and from the unprotected horizontal surfaces soaks through the roof structure (thermal insulation layer). The dampness then penetrates under the vapour barrier and into the floors of adjacent rooms, and from the bearing joint of the container flows down the vertical edge of the roof slabs, resulting in the decomposition and soiling of the structural members below.

Points for consideration

- On the upper structural surface of parapets a permanent watertight connection of the angled sealing layer ends by means of bonding; clamp rails or cement are not possible. This connection must be protected from water.

- Plant containers are not normally fully watertight in the long term; they let in rainwater through their sides or bottoms and into the bearing joint.

- Precast units such as plant containers and parapets are provided with open joints. These expansion joints are difficult to construct to be permanently watertight. If the sealing layer ends in front of such units, the roof slab and the roof structure in their supporting area have no protection against dampness.

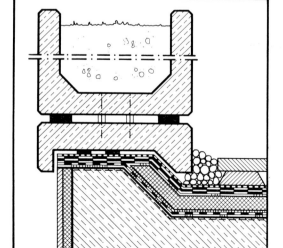

Recommendations for the avoidance of defects

● The sealing layer must be taken to the edge of the structural member and be carried to the required height of ⩾ 100 mm.

● If the parapet units at the edge of the structural slab (e.g. plant containers) are not permanently watertight, or have open joints which are difficult to seal, the sealing layer must pass below these units and be taken to the edge of the structural slab.

● The sealing layer may be connected to the parapet units only if these are permanently sealed against rainwater. The sealing layer must be then taken to a height of 150 mm and the sealing edge fixed and protected (see C 2.1.2).

Roof terraces
Edge of structural member

On roof terraces having edges formed from pressed metal sections bonded into the sealing layer, the fascias may warp and buckle, expansion joints open and soldered seams crack. Rainwater penetrates the defective sections and leads to the soiling and eventual corrosion of the fascias and damage to the parts of the structure lying below. In the region of the joints and the cracked soldered seams the firmly bonded sealing layer is damaged and lets in water. This causes saturation of the roof structure (thermal insulation), and the rooms below suffer.

At the eaves, leakages in the pressed metal gutters and cracks in the bonded sealing layer cause extensive water penetration. Near the edge, if sealing layers are constructed with little or no fall, ponding occurs and consequently the surface material becomes saturated. Distortion is seen on pressed metal fascias which warp and are unsupported.

Points for consideration

– The use of metal fascia sections to form the end of the sealing layer at the edge of the structural slab requires the direct connection of very different materials by means of an adhesive.

– These sections are not protected from sunlight and are exposed to a wide range of temperatures. With metal sections considerable changes of length and form result, which excessively stress other frictionally connected parts of the structure and lead to failure through compression (warping) and tension (cracks).

– Expansion joints are necessary to reduce the effective linear expansion in metal fascias, but these are expensive to construct and conceal new weaknesses.

– Pressed metal gutters on roof terraces with little or no fall of the sealing layer can, through deformations caused by temperature or as a result of faulty construction, produce an opposite fall to the surface of the structural member. When this occurs it prevents the complete drainage of the roof and in winter there is the added problem of frost in the edge area.

Recommendations for the avoidance of defects

● A frictional connection of the sealing layer with pressed metal edge sections should be avoided. Metal parts should be used only as cover elements.

● The increase in height necessary in angled edges should be formed by the sealing layer itself on a slope preformed in the sealing base.

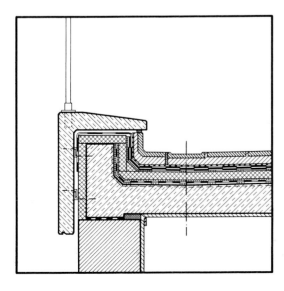

Roof terraces
Edge of structural member

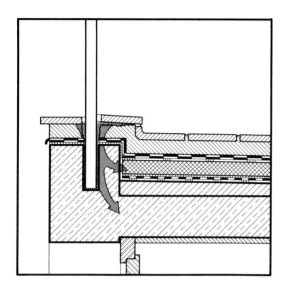

Balustrade supports which are fixed from above in the reinforced concrete roofing slabs and pierce the surface and roof structure cause, through leakages, extensive saturation of the roof structure and the interior. In some cases the pierced sealing layer is not connected to the balustrade post, or is connected only by grouting. In others, the connection becomes leaky when posts that are slender, or are secured only by a small baseplate to the roofing slab, move or bend. Near such posts, cracks occur in the surface and slabs loosen.

If the balustrade posts have been inserted from above in the reinforced concrete upstands – e.g. half-height parapets – then cracks may occur in the upstand, and the concrete spalls at the fixing points.

Points for consideration

- Balustrade posts fixed from above in the reinforced concrete roof slab pierce the surface and the roof structure, including the sealing layer, and defects quickly result.

- Through use and wind pressure balustrade posts are loaded horizontally. The flexible movements and bendings generated stress the sealing layer and surface finish. To avoid flexible movements of balustrading which is designed to stand on the surface of the structural slab, a fixing is required which is resistant to bending; with high balustrading, dimensions must be increased. The junction with the vapour barrier and the sealing layer must be made as independent as possible of the post by the use of flanges; the surface capping must be separated from the structural edge slab.

- Where balustrade posts are fixed directly to the edge upstand or slab, rainwater collects in the pockets that are unavoidably formed between the posts and the grouting; this leads to frost damage and spalling of the concrete and to corrosion of the posts.

Recommendations for the avoidance of defects

● Balustrades should be fixed outside the sealed surface of the roof terrace, in the loadbearing structural slab (e.g. front edge of reinforced concrete roof slabs).

● Balustrade fixings on the upper side of the roof slab, which pierce the surface and the sealing layer, should be avoided.

● If a fixing on the upper surface of the roof terrace cannot be avoided, the balustrade can be fixed in coping units – e.g. precast reinforced concrete units – which are laid on top of the sealing layer.

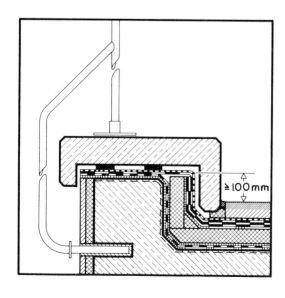

≥ 100 mm

Problem: Bearing surface and expansion joints

Roof terraces are, almost without exception, constructed on solid reinforced concrete roof slabs with thermal insulation on the upper side. The problems with the design of bearings and expansion joints are basically the same as those of flat roofs of solid construction.

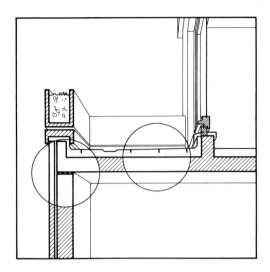

In comparison with flat roofs, however, roof terraces are usually considerably smaller. Because of this, the structural consideration of changes in form of the thermally protected roof slab is not necessary. In contrast, problems arise in the distribution and design of expansion joints in the terrace surface which are exposed to variations in temperature.

The most significant defects which have been found to occur in bearings and expansion joints on roof terraces are illustrated on the following pages, with recommendations for their avoidance.

Roof terraces

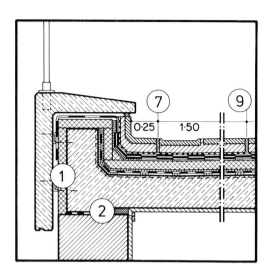

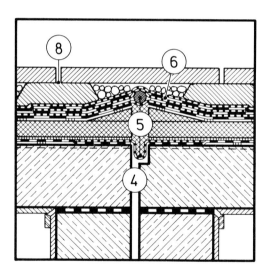

1 The edge beams and bearing of roof terraces should have an external thermal insulation layer with a resistivity to heat of approximately 1·3 m² K/W, which is directly connected to the thermal insulation layer of the terrace surface (see C 2.3.2).

2 If the roof terrace structural slab can have a firm bearing on the wall then an axial load should be ensured by inserting an approx. 10 mm thick strip of insulating material in the inner third of the bearing surface. Bearing and roof slab should be separated from one another by a foil (see C 2.3.2).

3 Where, in roof terraces with an interacting structural length longer than 12 m, the expected linear expansions are assimilated not by expansion joints but by displaceable bearings, in addition to the dowel reinforcement of the roofing slabs, dowels should be arranged over the loadbearing walls. Functional sliding bearings should be ensured by an absolutely smooth bearing joint (see C 2.3.2).

4 Structural separation at a width of about 20 mm and corresponding expansion joints, which reduce the interacting lengths of structural member, should be continued through all layers of the roof terrace (see C 2.3.3).

5 Where structural or expansion joints occur, the vapour barrier and sealing layer must be replaced by pliable foils or overlapping strips formed into a loop (see C 2.3.3).

6 The sealing layer should be raised above the drainage plane where this joint occurs (see C 2.3.3).

7 Roof terrace surfaces must be subdivided by expansion joints into areas with a maximum length of about 1·5 m and must be separated by an edge joint from all adjacent vertical members or those which pierce the surface (see C 2.3.4).

8 Where slabs are laid loose on gravel or on pad supports there must be open joints between the individual slabs. Their width is dependent on the dimensions of the slab, but must be ≥ 5 mm (see C 2.3.4).

9 Where flagstones or other surface finishes are laid in a mortar bed the surface expansion joints must continue at a width of ≥ 10 mm through the whole terrace surface structure to the separating layer. If these joints cannot be left open, because obstruction (e.g. by gravel) is expected, then this joint in the lower surface material should be filled with non-decaying, flexible material (e.g. strips of polystyrol foam) and be covered at the weathering surface by permanently flexible joint sections or mastic compounds (see C 2.3.4).

Roof terraces
Bearing surface and expansion joints

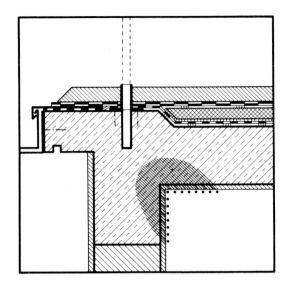

If roof terraces are not thermally insulated in the area of the bearing, the upper surface becomes damp and staining appears, especially at the corners of rooms below.

With roof terraces longer than 12 m, cracks appear in the edge beams and in the bearing brickwork.

Points for consideration

– The uninsulated bearing zone of roof terraces as well as uninsulated edge beams (to which balustrades are often fixed) represent thermal bridges which, internally in the corners of the room, can result in such low upper surface temperatures that surface condensation occurs.

– In designing for bending as well as linear expansion in the calculation of roof bearings, a roof terrace should be dealt with as a single slab solid roof (see A 1.1.3 – Linear expansion of roof slab, A 2.3.2 – Rigid bearing and A 2.3.3 – Movable bearing surface).

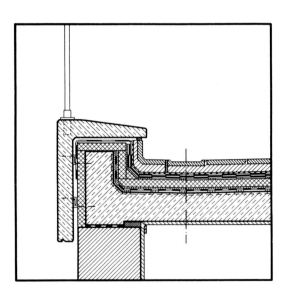

Recommendations for the avoidance of defects

● The edge beams and bearing of roof terraces should have an external thermal insulation layer with a resistivity to heat of about 1·3 m² K/W which is directly connected to the thermal insulation layer of the terrace surface.

● If the solid roof slab of roof terraces can be firmly mounted then a centrally located load (axial) on the wall should be ensured by inserting a strip of approx. 10 mm thick insulating material in the inner third of the bearing. Bearing and roof slab should be separated from one another by a foil.

● If in roof terraces with an interacting structural length of ⩾ 12 m the expected linear expansion is not assimilated by expansion joints but by a displaceable bearing, in addition to the dowel reinforcement of the roofing slabs, dowels should be arranged over the loadbearing walls and functional sliding bearings should be ensured by an absolutely smooth bearing joint.

Roof terraces
Bearing surface and expansion joints

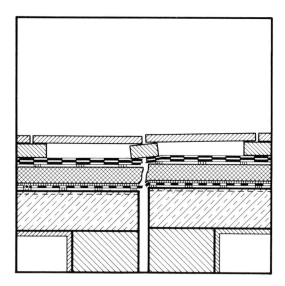

In the region of structural and expansion joints in the solid roof slab of roof terraces, failure of the surface layers and cracks in the sealing layer are found, which result in water penetration to the rooms inside.

On roof terraces with an interacting structural length longer than 12 m the lack of expansion joints or displaceable bearings results in cracks in the bearing area.

Points for consideration

— In order to reduce the interacting lengths of structural members, to separate the variously stressed areas and assimilate settlement, structural members are separated from each other by joints in the roof terrace. The weathering surface layers, especially the sealing layer, must absorb the variable movements without damage.

— An expansion joint in the sealing layer forms a complex constructional part of the roof membrane which, because it is stressed, is vulnerable to damage. Expansion joints should be sparingly used and should be raised above the level of standing water. Drainage over the expansion joint is not possible and its position and nature often decide the location and the number of necessary drainage outlets. The arrangement of the expansion joints must for this reason be considered at an early stage in design.

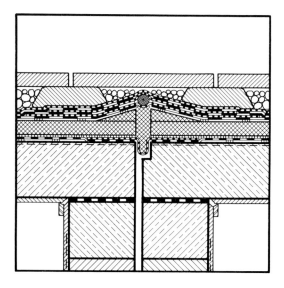

Recommendations for the avoidance of defects

● Structural separation at a width of about 20 mm and corresponding expansion joints, which reduce the interacting lengths of structural member, should be continued through all layers of the roof terrace.

● Where structural or expansion joints occur, the vapour barrier and sealing layer must be replaced by pliable foils or overlapping strips formed into a loop.

● The sealing layer should be raised above the drainage plane where such a joint occurs.

Roof terraces
Bearing surface and expansion joints

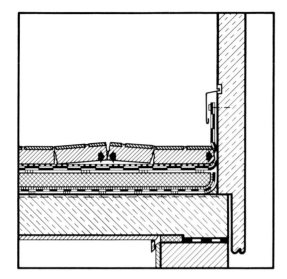

The surface of a roof terrace finished with slabs bedded in mortar, or with a cement or asphalt surfacing, frequently cracks and warps. This results in blistering of the surface and unevennesses, with cracks in the sealing layer and damp penetration. In all examples of such damage it has been found that the roof surface is not subdivided over a length of several metres and is firmly connected at the edge of the structural members.

Slabs laid on raised pads without adequate joint widths damage the upstand edge of the sealing layer and cause leakages.

Points for consideration

– Surface layers are subject to large variations in temperature and considerable linear expansion takes place. If movement is restricted, pressure stresses occur which can be absorbed only by a multi-layer construction, and to a limited extent this avoids cracking or warping.

– Defects in the surface or the sealing layer due to linear expansion are controlled if the surface layer is not firmly bonded to the base (see C 1.1.6 – Stressing of surfacing) or is separated from all vertical 'fixed points'. Alternatively, each individual surface slab or small area of surface can allow for certain movement within its own designed limits.

– The construction of expansion joints is simple in so far as these have no sealing function. If mastic joint fillers are absolutely necessary, filling with an elastic material should prevent the joint becoming clogged and free linear expansion of the surface can take place.

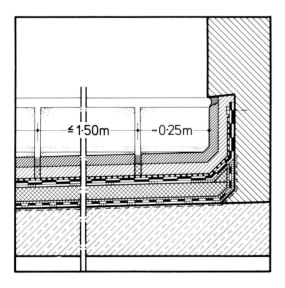

≤1·50m   ~0·25m

Recommendations for the avoidance of defects

● Roof terrace surfaces must be subdivided by expansion joints into smaller areas with a maximum length of approximately 1·5 m and must be separated by an edge joint from all adjacent vertical parts of the structure or those piercing the surfacing.

● Where slabs are laid loose on gravel or on support pads, open joints must be provided between the individual slabs; the width of these joints is dependent on the slab dimensions but must be ≥ 5 mm.

● Where flagstones or other surface finishes are laid in a mortar bed, the surface expansion joints must continue at a width of ≥ 10 mm through the whole terrace surface structure to the separating layer. If it cannot be left open because obstruction (by gravel) is expected, then this joint in the lower surface material should be filled with a non-decaying flexible material (e.g. strips of polystyrol foam) and be covered at the weathering surface by permanently flexible joint profiles or mastic compounds.

129

Problem: Drainage

The drainage of roof terraces demands the same care in respect of design and construction as that of flat roofs with no traffic. Similar sources of defects – e.g. non-watertight connections, construction of uninsulated outlets or outlets incorrectly placed, inadequate fall to the outlets – are to be expected. These lead to leakages and dampness of the roof terrace structure.

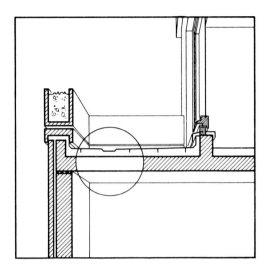

The construction of the outlet and the necessity for a watertight connection to the appropriate flange or foil results from the requirement of draining several layers (weathering surface – sealing – vapour barrier). Defects are frequently found here, with resulting water penetration damage which is compounded by the necessarily shallow fall. In some cases connections to the sealing layer are not provided, or are constructed incorrectly, so that the rainwater drainage does not function and the seeping water penetrates the surface.

Frequently recurring failures, resulting from faulty design detail as well as construction technique, and measures for their avoidance, are described in the following pages.

Roof terraces

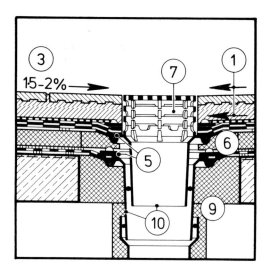

1 On roof terraces the rainwater from the roof surface as well as the seeping water from the sealing layer must be drained away by suitable drainage elements (see C 2.4.2).

2 External gutters should be used only on small roof terraces (see C 2.4.2).

3 With internally drained roof terraces the rainwater must be ducted away by means of an effective fall (1·5–2%) to outlets arranged at the lowest points (see C 2.4.2).

4 Roof terraces, because of their shallow fall, should be drained by means of several outlets (at least two) (see C 2.4.2).

5 Basically, every water-bearing layer should be connected to the drainage elements with a fall of 1·5–2%. Sealing layer and vapour barrier should be bonded to the outlet flange with a sufficiently large flexible connection foil or be connected with a multi-part fixed and loose flange (see C 2.4.3).

6 Reinforcement in the area of the outlet flange must be allowed for in the base (thermal insulation and concrete roof) (see C 2.4.3).

7 The outlet grating should be raised to the height of the surface either with the aid of adaptor rings or, in the case of pad-supported surfaces, by fixing to the pads (mortar dabs) (see C 2.4.3).

8 Where slabs are laid loose on gravel, washed gravel of a particle diameter ⩾ 16 mm should be used (see C 2.4.3).

9 Outlets and drainpipes must be thermally insulated (see C 2.4.3).

10 As a general principle, multi-part outlet elements should be installed. They should be adjustable to suit the various structural heights and should be protected from backwash water by means of a roller cage and lip sealing (see C 2.4.3).

11 Roof outlets must be constructed so that they do not shift under loading. They must, equally, be accurately located in the concrete roof (see C 2.4.3).

Roof terraces
Drainage

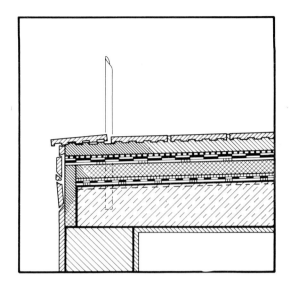

Roof terraces of which the only possible means of drainage is the overflow of rainwater over the roof edge exhibit blistering and spalling of the cladding (plaster, tiles) at the roof edge together with a damp appearance of the complete wall below the edge of the roof terrace, and sometimes even inside the building.

Roof terraces drained by external gutters, as well as defective construction (the edge trim having an inward fall) sometimes show defective bonding between gutter fascia and sealing panel, so that water can penetrate beneath the gutter fascia and behind the rendering, which later spalls off.

Puddles form on the surface and sealing layers of inwardly drained roof terraces if only one outlet on the narrow side is provided to drain the surface. Similarly, ponding results when the outlet elements drain only the upper surface of the terrace, or where the fall to the outlets is too shallow.

Points for consideration

— On roof terraces where the rainwater can drain away only over the terrace edge, the flow of water, increased by wind pressure, may run underneath the slab surface and the sealing layer.

— Where the front edge of the terrace is finished with rendering or slabs, the water cannot flow away unhindered on the sealing layer; water accumulates behind the cladding, causing voids, blistering and chipping or spalling as a result of frost action.

— External gutters must be bonded with their eaves' fascias in the sealing layer. This is very difficult to achieve in long roof terraces because of the necessary arrangement of expansion joints.

— Individual outlets and gutters can drain rainwater only from that part of the roof surface to which they are connected by a fall. Ponding on the surface and on the sealing layer detracts from the terrace's usefulness and may lead to further damage.

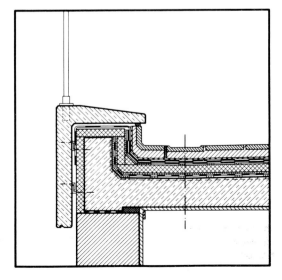

Recommendations for the avoidance of defects

● The rainwater must be drained from the upper surface of roof terraces, and the seeping water from the sealing plane, by means of drainage elements.

● Where roof terraces drain inwardly, the rainwater must be ducted to outlets arranged in the depression by means of an effective fall (1·5–2%).

● Because of the shallow fall, at least two outlets should be provided.

● External gutters should be used only in small-surface roof terraces.

Roof terraces
Drainage

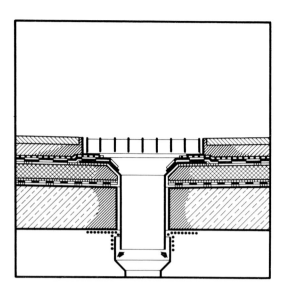

Construction defects in the connection to the layers to be drained occur in internal outlet elements (gullies) as well as external gutters. Sometimes, due to underestimation of the function of the sealing layer, these layers are quite unconnected, or are incorrectly connected, to the drainage units. The water penetrating through the upper surface to the sealing layer cannot flow freely and remains in puddles beneath the surface. Following frost damage, defects appear in the surface and dampness penetrates to the structure below.

Where slabs are laid on a gravel bed drainage becomes blocked, as fine particles of the gravel base settle in front of the outlet filters and prevent dispersal of the water. As in the flat roof, badly constructed outlet elements are often installed. If the drainage elements – outlets and drainpipes – are uninsulated or are constructed without backwash protection, water penetrates the whole structure in the area of these drains.

Points for consideration

- The roof terrace has two water-bearing planes: the upper surface of the terrace and the sealing layer. Both must be connected to the drainage elements. In pad-supported surfaces with open joints the rainwater runs on the plane of the sealing layer.

- If slabs are laid on loose gravel, washing-out of fine particles and resultant blocking of outlets must be prevented by specifying a suitable size of gravel particle.

- Drainage of the reinforced concrete roof slab during the initial construction period should be possible.

- The outlet elements must be adjustable to the various structural heights of the surface and thermal insulation layers.

- If multi-part outlet elements are necessary for drainage, no water should be allowed to get into the roof structure through backwash.

- If the roof structure is loaded from above, there is a danger of outlets tilting or loosening, causing damage to the sealing layer and thermal insulation.

Recommendations for the avoidance of defects

● Every water-bearing layer must be connected to the drainage elements by a fall of 1·5–2%. Sealing layer and vapour barrier should be bonded to the outlet flange with a sufficiently large flexible connection foil or joined with a multi-part fixed and loose flange.

● The increased thickness of the sealing layer in the area of the outlet flange must be allowed for in the base (thermal insulation and roofing).

● The outlet grating should be taken to the height of the surface finish either with the aid of adaptor rings or, in pad-supported slab surfaces, by fixing on the pads (mortar dabs).

● Where slabs are laid on loose gravel, washed granules of diameter ⩾ 16 mm should be used.

● Outlets and drainpipes must be thermally insulated.

● In principle, multi-part outlet elements should be installed. They should be adjustable to the variable structural height and should be protected from backwash water by a cage and lip seal.

● Roof outlets must be constructed in such a way that they remain firm in their original position after loading. They must, correspondingly, be accurately located in the concrete roof.

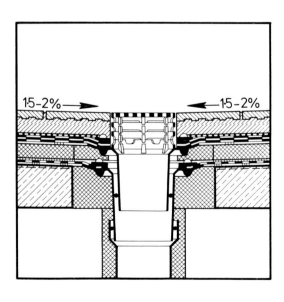

Problem: Thresholds

The design of roof terraces, i.e. usable roof surfaces over internal rooms, necessitates access from an internal room to outside, usually through a storey-height glazed door.

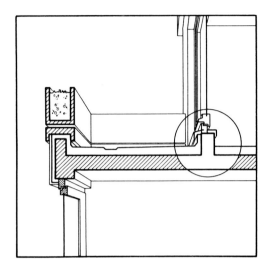

The result of the research investigations into building defects showed that this access over the door threshold presents a particular problem with roof terraces. One in two of all observed defects on roof terraces occurs at this point. Analysis shows recurring failure in design or construction as the cause of damp penetration to the internal floors or roof structure and the rooms below. Primarily, the same defects are found as are observed at the junction to vertical parts of the structure (see D 2.1). However, defects in the sealing layer upstand also seriously affect the threshold area.

Roof terraces

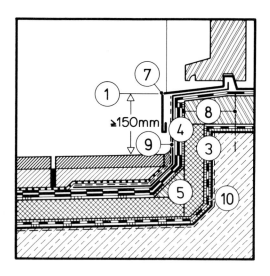

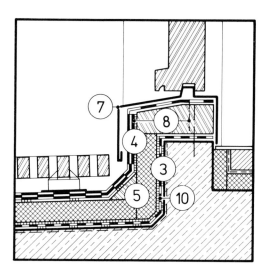

1 At the thresholds of doors and windows the sealing layer must be raised above the highest water level that may be expected in adverse conditions. The top of the sealing layer must therefore extend ⩾ 150 mm above the upper surface of the roof terrace (see C 2.5.2).

2 If a high threshold is not possible for reasons of traffic (e.g. wheelchairs), the transition from the roof terrace surface to the inner room must be designed to preclude rainwater. This can be achieved by recessing the door behind the outer face of the wall and by providing a steep fall to the rest of the roof terrace surface (see C 2.5.2).

3 The structural height of the sealing edge, which is calculated from the height of the sealing upstand and the number and thickness of individual layers – surface slab, roof structure, falls and underlays – is known at design stage and determines the threshold height (see C 2.5.2).

4 The angle and the junction of the sealing layer to the threshold should be formed from the roof membrane itself or by the use of foils of similar material behaviour (see C 2.5.4).

5 The transition of the sealing layer to the angle should be as uniform as possible, avoiding sharp bends at change of plane. The gradual transition must be preformed in the sealing base, e.g. by triangular wedges made of thermal insulating material, ensuring a continuous sealing base (see C 2.5.5).

6 The use of metal sections to form the junction of the door threshold should be avoided in threshold lengths > 3 m (see C 2.5.4).

7 The edge of the sealing layer at the threshold must be protected from the water running down the door and window surfaces. This is normally achieved by inserting a pressed metal cover section or window sill over the sealing edge layer (see C 2.5.3).

8 To avoid slipping or loosening, the sealing layer must be continuously fixed along its entire length, preferably by a clamp rail (see C 2.5.3).

9 The angled sealing layer must be protected from excessively strong thermal or mechanical stresses by the coving of the weathering surface slabs or by sheet cladding suspended in front of it (see C 2.5.4).

10 The roof structure above the loadbearing slab, especially the thermal insulation layer, must be additionally protected at the lateral surface against possible damp penetration by continuing the vapour barrier all round and fixing it with adhesive (see C 2.5.3).

Roof terraces
Thresholds

Defective threshold construction usually results in serious dampness in the roof structure (thermal insulation layer) and in the internal floors or the rooms below. In some cases the sealing layer is not formed into an upstand, or is not raised high enough above the upper surface of the roof terrace. Although the threshold may be raised above the internal floor level, provision of a considerable angle of fall outside means that inadequate upstand height is available, since at the threshold the paving slabs are at the same level as the floor inside. Moreover, in some cases the surface slab is laid without a fall or even with a fall towards the threshold, so that driving rain, wind-banked water and melting snow can overflow the sealing layer or the threshold.

Points for consideration

— Like all junctions of the sealing layer to vertical parts of the structure, the threshold connection is subjected in certain circumstances to adverse stresses from melted snow, banked-up water and spray. Comparable conditions are found only in the region of the building foundation.

— The structural depth of the roof terrace structure, because of thermal insulation and falls, is usually greater than the structural depth of the internal floors.

— The difference of height between surface slabs and threshold can be achieved by cranking in the roofing slab or by a threshold slope.

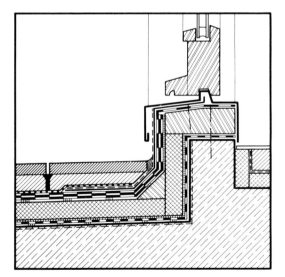

Recommendations for the avoidance of defects

● At the threshold of windows and doors the sealing layer must be raised above the highest possible level of water that may be expected in adverse conditions. The end of the sealing layer must extend ⩾ 150 mm above the upper surface of the roof terrace.

● If a high threshold is not possible for reasons of access (e.g. wheelchair), the transition from the interior to roof terrace surface must be protected from rainwater penetration. This can be achieved by recessing the door behind the outer face of the wall and by providing a steep fall to the rest of the roof terrace surface.

● The structural height of the sealing edge, which is calculated from the height of the sealing upstand and the number and thickness of individual layers – surface slab, roof structure, falls and underlays – is known at design stage and determines the threshold height.

Roof terraces
Thresholds

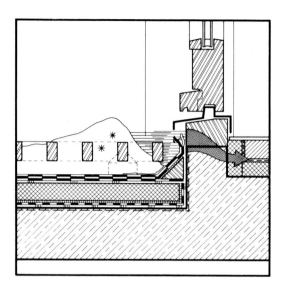

In some cases, although the sealing layer is raised at the threshold, rainwater penetrates behind its edge. The edge of the sealing layer may be insufficiently covered by the metal flashing, or the edge of the sealing angle may be inadequately fixed to the threshold, come loose and slip down.

Sealing edges which are sealed with cement leak. If the roof structure (thermal insulation layer) is not sealed at the lateral surface, in addition to damp penetration of internal floors and the rooms beneath, the roof structure becomes unserviceable as a result of extensive dampness.

Points for consideration

- During a rainstorm the external face of the building is subjected to a large quantity of water on its surface, and with high winds this is driven into all joints and cracks in the weathering surface; it can even penetrate over the edge of sealing membranes if these are not adequately protected.

- In practice, the junction of the sealing layer to the threshold cannot be made permanently watertight. Sealing by means of cement has limited durability and requires constant maintenance.

- The roof structure, especially the thermal insulation, can in certain circumstances be so damaged through dampness that it must be replaced.

Recommendations for the avoidance of defects

● The lip of the sealing layer at the threshold must be protected from water running down the impervious surfaces of the windows and doors. This is achieved by inserting the sealing layer under the metal threshold plate or window sill.

● To avoid slipping or loosening, the end of the sealing layer must be evenly fixed along its whole length, preferably by means of a clamp rail.

● The roof structure above the loadbearing slab, especially the thermal insulation layer, must be further protected at the lateral surface from damp penetration, e.g. by continuing the vapour barrier all round and fixing it with adhesive.

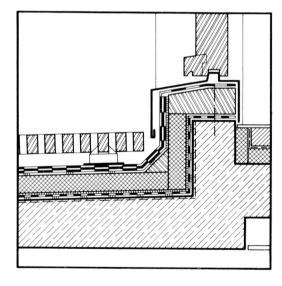

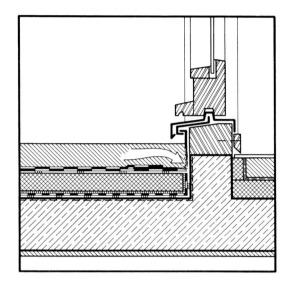

Where metal sections are used to form the junction between the sealing layer and the vertical threshold upstand, many leakages result. In some cases, the connection of the metal section to the roofing panels is not watertight because the metal section has not been carefully bonded between the thicknesses of the sealing layer or has been bonded only on one side. In other examples, sections made of lead and zinc bedded in the mortar bed of the roof surface slabs have been designed without a protective layer; these fail and leak as a result of corrosive attack by the mortar particles.

Points for consideration

— The use of sheet metal upstands requires the permanent watertight bonding of two very different structural materials with the aid of an adhesive; this demands special care.

— Metals, especially milled lead and zinc, are vulnerable to damage by lime or cement mortar in that the calcium hydroxide $(Ca(OH)_2)$ dissolves the metal. A direct contact must be avoided and the metals provided with a protective finish.

— Sheet metal upstands are subjected to strong changes in temperature and undergo corresponding linear expansion.

Recommendations for the avoidance of defects

● The upstand and junction of the sealing layer to the threshold should be formed from the roof membrane itself or by using connection foils of similar material properties.

● The upstand sealing layer must be protected against excessive thermal or mechanical stress by the coving of the weathering surface slab or by means of freely moving sheet metal cladding protection.

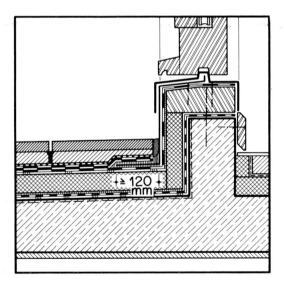

● The construction of the threshold junction using sheet metal should be avoided where the threshold length is > 3 m.

● If the use of metal sections is unavoidable, they should be protected against the destructive action of lime and cement mortar by separating layers or protective coatings (e.g. bitumen). Sliding seams should be provided, to allow for linear movement, at adequately spaced centres (with titanium zinc, every 3 m). After the removal of dampness and dirt (oil, etc.) the metal sections should be inserted ≥ 120 mm between the individual layers of the multi-layered sealing material.

Roof terraces
Thresholds

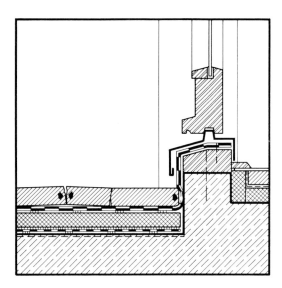

At solid threshold upstands leakages occur near the edge. Where surface slabs are laid directly against the upstand (flagstones, pad-supported slabs) deformation and damage to the sealing layers result (see C 1.1.7). Where the sealing layer is deflected into a right angle and unsupported, it is pierced during use of the surface of the structural slab in the construction period.

Sealing edges which are only bonded to the threshold or fixed at intervals by nailing become loose and rainwater penetrates behind them (see C 2.2.5).

Points for consideration

– During the construction period the unprotected sealing layer undergoes mechanical stresses from foot traffic, passage of wheelbarrows, storing of construction materials, etc.

– Multi-layer sealing layers and foils cannot be deflected into a right angle. At sharp angles the bituminous bonding layer therefore cracks and the sealing layer is unsupported.

– The vulnerability of the sealing layer to mechanical damage is increased in areas where it is unsupported (see C 1.1.5 – Stressing and protection of sealing layer).

– The weathering surface layers are exposed to extreme changes in temperature and are subjected to strong thermal linear expansion. As the surface slabs are loosely applied on the sealing layer, this linear expansion takes particular effect at the edge of the continuous surface, e.g. at the junction with a wall surface where the sealing layer is angled (see C 1.1.6 and 2.3.4).

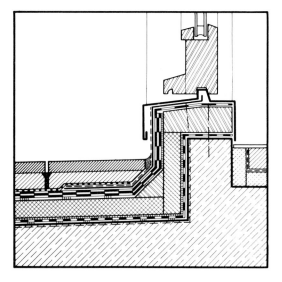

Recommendations for the avoidance of defects

● The transition of the sealing layer to the upstand should be as uniform as possible, avoiding acute angles at changes of plane; the gradual transition must be preformed in the sealing base, e.g. by triangular wedges made of thermal insulating material, ensuring a continuous sealing support.

● The surface finish of roof terraces up to the separating layer must be segregated from the upstand by an edge joint which effectively prevents the stressing of the angled sealing layer by the thermal expansion of the surface finish.

General texts and principles

Eichler, F.: Bauphysikalische Entwurfslehre, Band 2, 4. Auflage, Verlagsgesellschaft Rudolf Müller, Köln 1973.

Moritz, Karl: Flachdachhandbuch – Flache und flachgeneigte Dächer, 4. Auflage, Bauverlag Wiesbaden und Berlin 1975.

DIN 1045 – Beton- und Stahlbetonbau, Januar 1972.

DIN 1053 – Mauerwerke, Berechnung und Ausführung, Blatt 1, November 1974.

DIN 18530 – Massive Deckenkonstruktionen für Dächer, Dezember 1974.

DIN 4122 – Abdichtung von Bauwerken gegen nicht drückendes Oberflächenwasser und Sickerwasser mit bituminösen Stoffen, Metallbändern und Kunststoffolien; Richtlinien, Juli 1968.

Connection to vertical abutments

Haefner, Rudolf: Abdichtungsprobleme beim Bau der Parkdecke des Rhein-Neckar-Hochhauses in Mannheim. In: Abdichtung von Ingenieurbauwerken, Band 4 der Schriftenreihe der Bundesfachabteilung Abdichtung gegen Feuchtigkeit, Bauverlag Wiesbaden und Berlin 1967.

Hoch, Eberhard: Flachdächer mit harter Schale – Terrassendächer und Parkdeckabdichtungen. In: Das Dachdeckerhandwerk (DDH), Heft 11/72, S. 788–792.

Kakrow, Helmut: Parkdeck und Hofkellerabdichtungen. In: Deutsches Dachdeckerhandwerk (DDH), Heft 9/69, S. 506–515.

Lufsky, Karl: Bauwerksabdichtung – Bitumen und Kunststoffe in der Abdichtungstechnik, 2. Auflage, B. G. Teubner, Stuttgart 1970.

Rick, Anton W.: Risse und Randabsetzungen bei Asphaltterrassenbelägen. In: Bitumen, Teere, Asphalte, Peche…, Heft 10/73, S. 422–423.

Zimmermann, Günter: Dachterrassen mit aufgesetzten Belägen – Durchfeuchtungen von Decken und Wänden. In: Deutsches Architektenblatt (DAB), Heft 12/74, S. 927.

Edge of structural member

Haefner, Rudolf: Abdichtungsprobleme beim Bau der Parkdecke des Rhein-Neckar-Hochhauses in Mannheim. In: Abdichtung von Ingenieurbauwerken, Band 4 der Schriftenreihe der Bundesfachabteilung Abdichtung gegen Feuchtigkeit, Bauverlag Wiesbaden und Berlin 1967.

Hoch, Eberhard: Flachdächer mit harter Schale – Terrassendächer und Parkdeckabdichtung. In: Das Dachdeckerhandwerk (DDH), Heft 11/72, S. 788–792.

Kakrow, Helmut: Parkdeck und Hofkellerabdichtungen. In: Deutsches Dachdeckerhandwerk (DDH), Heft 9/69, S. 506–515.

Probst, Raimund: Außengänge, Balkone, Dachterrassen. In: Baupraxis, Heft 3/70, S. 39–43.

Timmerberg, Carl H.: Details … Details, Baupraktische Hinweise zu Esser-Produkten – Essers kleine Handbuchreihe, Herausgeber: Klaus Esser KG, Düsseldorf 1974.

Bearing surface and expansion joints

Balkowski, F. D.: Die Rißbildung am Deckenauflager. In: Das Dachdeckerhandwerk (DDH), Heft 2/74, S. 88–91.

Brandes, K.: Dächer mit massiven Deckenkonstruktionen – Ursache für das Auftreten von Schäden und deren Verhinderung. In: Berichte aus der Bauforschung, Heft 87, Verlag Wilhelm Ernst & Sohn, Berlin 1973.

Engels, Friedhelm: Bauwerksabdichtungen – begehbar – befahrbar – bepflanzt. In: Deutsches Dachdeckerhandwerk (DDH), Heft 10/68, S. 404–606.

Kramer-Doblander, Herbert: Temperaturspannungen in Flachdachkonstruktionen. In: Bitumen, Teere, Asphalte, Peche…, Heft 3/70, 21. Jahrgang, S. 93–96.

Lufsky, Karl; Konzack, Kurt: Begehbare Dachterrassen. In: Bitumen, Teere, Asphalte, Peche…, Heft 10/67, S. 359–368.

Pfefferkorn, Werner: Konstruktive Planungsgrundsätze für Dachdecken und ihre Unterkonstruktionen. In: Das Baugewerbe, Heft 18/73, S. 57–65; Heft 19/73, S. 54–59; Heft 20/73, S. 86–90; Heft 21/73, S. 54–63.

Rick, Anton W.: Einiges über Asphaltbeläge auf Terrassen. In: Bitumen, Teere, Asphalte, Peche…, Heft 2/72, S. 82–83.

Rick, Anton W.: Risse und Randabsetzungen bei Asphaltterrassenbelägen. In: Bitumen, Teere, Asphalte, Peche…, Heft 10/73, S. 422–423.

Schütze, Wilhelm: Der Estrich auf Dächern und Terrassen, eine Estrichart, die ganz besondere Sorgfalt erfordert. In: Boden, Wand und Decke, Heft 3/65, S. 182–192; Heft 4/65, S. 294–304.

Ullmann, W.: Terrassen – begehbar – befahrbar – bepflanzt. In: Das Dachdeckerhandwerk (DDH), Heft 11/72, S. 797–801.

Zimmermann, Günter: Wärmedämmung bei nicht belüfteten Flachdächern. In: Das Dachdeckerhandwerk (DDH), Heft 10/68, S. 597–602.

AGI-Arbeitsblätter; A 10 – Industrieböden, Hartbetonbeläge; A 11 – Industrieböden, Zementestrich als Nutzboden.

Drainage

Buch, Werner: Das Flachdach, Dissertation, Darmstadt 1961.

Haefner, R.: Die Abdichtung von unterkellerten Hofdecken, Terrassen über Nutzräumen und Flachdächern. In: Böden, Wand und Decke, Heft 7/66, S. 588–696.

Hoch, E.: Flachdächer, Flachdachschäden, Verlagsgesellschaft Rudolf Müller, Köln 1973.

Österreichisches Institut für Bauforschung: Abdichtungen und Abläufe bei Flachdächern, Dachterrassen, Balkonen, Loggien, Naßräumen, 2. Auflage, Wien 1973.

Probst, Raimund: Außengänge, Balkone, Dachterrassen. In: Baupraxis, Heft 3/70, S. 39–43.

Timmerberg, Carl H.: Details … Details…, Baupraktische Hinweise zu Esser-Produkten, Essers kleine Handbuchreihe, Herausgeber: Klaus Esser KG, Düsseldorf 1974.

Voorgang, H. J.: Die Entwässerung flacher Dächer. In: Deutsches Dachdeckerhandwerk (DDH), Heft 1/69, S. 130–132; Heft 6/69, S. 278–282; Heft 9/69, S. 532–535.

DIN 1986 – Grundstücksentwässerungsanlagen, Juni 1962.

Thresholds

Lufsky, Karl: Bauwerksabdichtung – Bitumen und Kunststoffe in der Abdichtungstechnik, 2. Auflage, B. G. Teubner, Stuttgart 1970.

Probst, Raimund: Außengänge, Balkone, Dachterrassen. In: Baupraxis, Heft 3/70, S. 39–43.

Zentralverband des Dachdeckerhandwerks: Richtlinien für die Ausführung von Flachdächern, Ausgabe Januar 1973, Helmut Gros Fachverlag, Berlin 1973.

Problem: Sequence of layers and individual layers

The balcony provides a private external area to a dwelling, and forms an integral part of the basic layout of the living unit in many blocks of flats. However, lack of consideration of orientation and size frequently reduce the value of the balcony in its function and as a design element of the façade.

Defects of construction frequently occur, resulting either from over-stressing of the balcony as an external structural member, or from the structural disregard of the balcony. The basic recommendation for the avoidance of defects is not to have balconies unless they can be constructed as useful free surfaces and therefore increase the value of the dwelling.

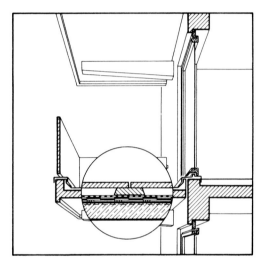

In general, damage to balconies, loggias and arcades – which have the same structural problems – is given little attention provided there is no acute danger of their collapsing and significantly affecting the interior. This may be due partly to the minimal demands which are made of balconies, as areas outside the dwelling, in terms of absence of cracks, cleanness, nature of their surface, protection against damp; and partly to the attitude that defects in balconies (like those in cellars) are to be accepted as 'inevitable'.

In the recommendations on the following pages, which are derived from examples of defects to balconies with solid reinforced concrete slabs, balconies and arcades are regarded not as simple structural units of correspondingly simple construction, but as useful structural members which, like other areas of the building, must remain free from damage.

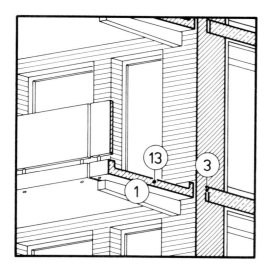

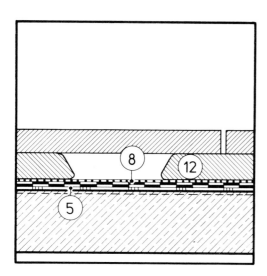

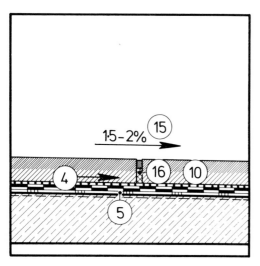

1 Balcony slabs should if possible be constructed of precast slabs displaceably mounted on side walls or cantilevers and separated from the adjoining floor slabs (see D 1.1.2).

2 If the balcony is constructed as a cantilevered slab, this must be subdivided by expansion joints at maximum centres of 4 m. These must separate existing solid parapet elements and continue up to the bearing wall (see D 1.1.2).

3 In order to avoid thermal bridges, the external face of the floor slab should be protected with thermal insulation. Where the balcony slab is cantilevered, the balcony area and the floor slab must be protected for a width of approximately 500 mm with a thermal insulation layer $\geqslant 20$ mm thick (see D 1.1.3).

4 Underneath the weathering surface layer there must be a rotproof sealing layer. This must have a fall of 1·5–2·0% to the outlet points (see D 1.1.4).

5 The lower layer of the sealing membrane should be strip or spot bonded to the loadbearing slab (see D 1.1.4).

6 The upper surface of the finished sealing layer should, because of its shallow fall, be free from unevennesses or protrusions (see D 1.1.4).

7 For the protection of the sealing layer the weathering surface should, if possible, be applied directly after the sealing work is completed (see D 1.1.4).

8 Between the sealing layer and the weathering surface a durable separating layer must be inserted; this normally consists of two layers of loose-laid polyethylene foil (see D 1.1.4 and 1.1.5).

9 The upper weathering surface layer should be constructed with water-permeable joints on layers which are as permeable as possible (see D 1.1.6).

10 Weathering surfaces consisting of flagstones or slabs in a mortar bed should be bedded on protective concrete with porous joints and a thickness of 50 mm (see D 1.1.6).

11 Weathering surfaces consisting of manufactured slabs of natural or artificial stone with dimensions of $\geqslant 400 \times 400$ mm are laid in a level precompressed gravel layer $\geqslant 50$ mm thick of washed granules 16–32 mm in diameter (see D 1.1.6).

12 Weathering surfaces of precast slabs made of natural or artificial stone can be laid on pads. If possible, wide surface bearings which will balance out any unevennesses (e.g. mortar spots) should be used as pad mountings and the slabs should be laid with minimal expansion joint widths (> 5 mm) (see D 1.1.5 and 1.1.6).

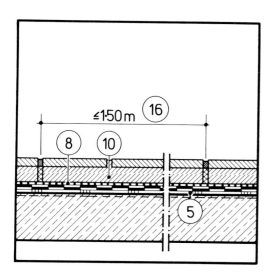

13 Precast reinforced concrete balcony slabs with a waterproof wearing surface, laid on cantilever bearings in front of the façade, in dimensions of up to approx. 4 m, can be installed as complete units without the need for additional sealing and surfacing layers (see D 1.1.4 and 1.1.6).

14 A poured asphalt finish should be laid only on surfaces that are protected from direct sunlight and are not subjected to point loads (see D 1.1.7).

15 The upper surface of the weathering surface should be laid to a fall of 1·5–2% and should be connected to the drainage. Where slabs are laid in a gravel layer or on pads, if necessary the fall and drainage of the upper surface finish can be omitted (see D 1.1.7).

16 The surface finish must be separated by separation joints from all adjacent structural members and from those which pierce it, and it must be subdivided by functional expansion joints into areas with a maximum length of approx. 1·5 m (see D 1.1.5).

Balconies
Sequence of layers and individual layers

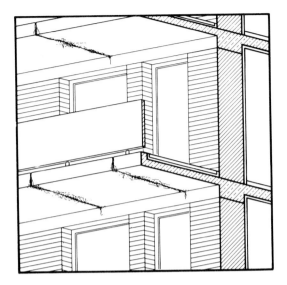

The most frequently occurring damage to balconies takes the form of cracks in the reinforced concrete slab, running at right angles to the front of the house. These are found in cantilevered slabs > 4 m in length. Often the sealing layer and the surface finish above these cracks also fracture. Soaking, erosion and efflorescence on the under-side of the slab are the result. In some cases the cracks continue into the floor at that level, resulting in damp penetration to the internal rooms.

Points for consideration

— While the variation in temperature of the loadbearing cross-section of externally insulated roof and wall structural members is modified by the relatively constant internal temperature, the balcony cross-section is fully exposed to variations of outside temperature and thus can be heated up by intensive sunlight. Similarly, balcony slabs are exposed on all sides to external variations of humidity.

— Wide variations in temperature and dampness cause large linear expansions which, when restricted, can lead to compressive or tensile stresses. Especially where lateral reinforcement is defective, these stresses can lead to cracks in reinforced concrete.

— Cantilevered slabs are restricted in their linear expansion by their firm connection to the adjoining floor which is exposed to the constant internal climate; they are therefore particularly vulnerable to damage.

— The thermal insulation of balcony slabs can reduce the effect only of short-term changes in temperature. Where temperature conditions are more long-lasting, thermal insulation of structural members exposed on all sides is ineffective as there is no counterbalancing effect of the constant internal climate. Thermal insulation on the upper side therefore has a very limited effect in strong sunlight; thermal insulation on the under-side does not reduce the temperature loading.

— The negative consequences of linear expansion of balcony slabs can be avoided by the prevention of stresses – by separation of storey floor and balcony slab – as well as by the reduction of the interacting lengths of structural member – by insertion of expansion joints.

Recommendations for the avoidance of defects

● Balcony slabs should if possible be constructed of precast slabs displaceably mounted on side walls or cantilevers and separated from the adjoining floor slabs.

● If the balcony is constructed as a cantilevered slab, it must be subdivided at maximum centres of 4 m with expansion joints which, if necessary, should continue through solid parapet elements and to the bearing wall. The stresses that arise must be absorbed by adequate lateral reinforcement.

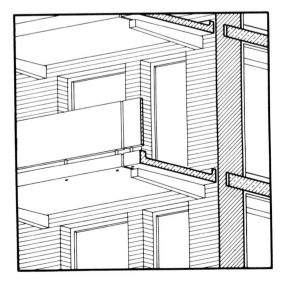

Balconies
Sequence of layers and individual layers

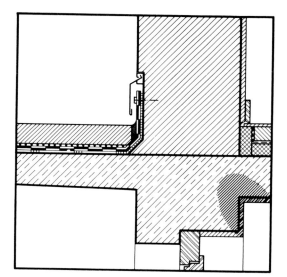

Cantilevered balcony slabs which are firmly fixed to the adjoining floor draw out the warmth. In unfavourable circumstances the result is condensation and staining in the wall and ceiling areas below the roof slab, and sometimes even in the skirting area above.

Points for consideration

– Because of the high thermal conductivity of reinforced concrete and the considerable cooling surface of the balcony exposed to the external climate, reinforced concrete cantilever slabs act as thermal bridges.

– Even where upper surface condensation in the region of the thermal bridge does not normally occur, because the convection flow of the room heating on the window side maintains above condensation point the upper surface temperature in the ceiling area above the window, allowance must be made for possible drops in temperature and the danger of upper surface condensation.

– Because of its large external surfaces, external insulation of the cantilever slab in respect of the loss of heat from the interior is completely ineffective. The thermal bridge effect in cantilever slabs can be reduced only by internal insulation of the front wall and ceiling area below.

– The separation of floor and balcony slabs permits simple and problem-free front surface insulation of the roofing slab.

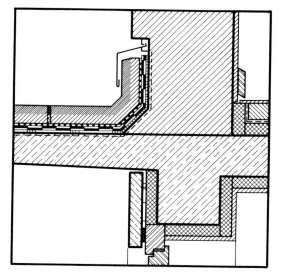

Recommendations for the avoidance of defects

● Balcony slabs should if possible be constructed separately from the adjoining slab on side walls or cantilevers. In order to avoid a thermal bridge the adjoining slab should be provided with insulation to the front surface.

● If balcony slabs are constructed as cantilevers, the front wall below, and the adjoining slab for a 500 mm width, should be provided with a thermal insulation layer $\geqslant 20$ mm thick.

Balconies
Sequence of layers and individual layers

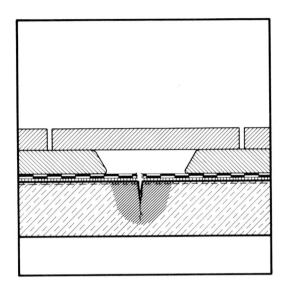

With balconies, loggias and arcades, one frequently finds damp areas caused by rainwater penetrating the cantilevered slabs and sometimes also connected parts of the structure: beneath weathering surfaces which have been assumed to be waterproof the sealing layers are completely lacking or are incorrectly laid. Often the sealing layer is not continued to the edge of the structural member (parapet or vertical wall) and properly secured.

Together with erosion and freezing of the surface layers due to ponding on the sealing layer, the following examples of damage may be found in the sealing layer itself:

– signs of decay where non-waterproof sealing panels are used;

– cracks caused by frictional connection of the sealing layer either with the loadbearing slab or with the completely bonded upper surface finish;

– holes resulting from mechanical stresses caused by flagstone or precast slab installers working on the unprotected sealing layer.

Points for consideration

– Although because of its protected position only a limited quantity of rainwater has to be disposed of, the sealing layer of a balcony, loggia or arcade, like that of a flat roof, has to protect the loadbearing slab from penetration by surface water and to conduct water which penetrates the surface to the drainage elements as quickly as possible. Weathering surface layers, whether flagstone or cold asphalt, cannot fulfil a sealing function (see D 1.1.6 – Design of surface finish).

– Control of the functioning of the sealing layer is difficult to achieve: great care is needed not only in the physical operation of laying the sealing layer, but also in the selection of suitable materials.

– Ponding on the sealing layer increases the danger of penetration into the base structure, especially where the bonding of overlap joints is defective.

– Ponding on the sealing layer leads to damage of the surface finish through erosion, efflorescence and freezing (see D 1.1.6 – Design of surface finish).

– The build-up necessitated by bonding of the sealing layer to the flanges of the outlet elements prevents the water from flowing away completely.

– Sealing panels with felt or jute fabric linings are liable, under constant wetness, to decay and thereby to leakages.

– Sealing layers bonded completely on to the base or frictionally connected to the upper surface are totally exposed to the movements of the adjacent layers and are strongly stressed by tension and expansion.

– The inclusion of dampness and air under the sealing layer, generated by a wet and dirty base, leads in intense heat under pad-mounted slab surfaces to excess pressure and, where there is inadequate provision for the absorption of stress, to the formation of bubbles.

– Sealing layers, whether bituminous or of synthetic foils, are extremely vulnerable to mechanical stresses, especially where there are hollow places under the sealing layer caused by unevennesses in the base.

– With regard to the stressing of the sealing layer by pad-mounted surfaces, the criteria mentioned in respect of roof terraces apply here (see C 1.1.5 – Stressing and protection of sealing layer).

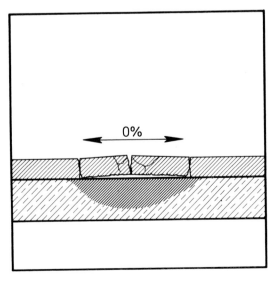

0%

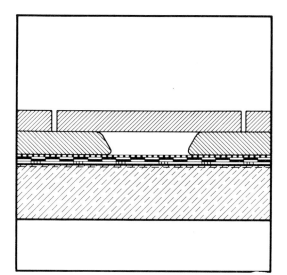

— When waterproof reinforced concrete precast slabs are used in balcony construction, supported on cantilevers and independent of the internal floor slab, they do not require further sealing or surface layers provided they remain free of cracks (see D 1.1.6 – Design of surface finish).

Recommendations for the avoidance of defects

● The sealing layer must have a fall of 1·5–2% to the outlets. Where falls are greater, the surface finish must be secured from slipping by abutments or adhesion.

● The recommendations made for flat roofs apply to the method of laying of the sealing layer (see A 1.1.9 – Laying of sealing layer).

● Where the balcony fall is shallow, care should be taken to obtain a flat finished upper surface of the sealing layer with no protuberances.

● Roofing panels with linings which are not stable in damp conditions (felt and jute fabric) should not be used.

● The first layer of the sealing membrane should be spot or strip bonded to the loadbearing slab, using a glass-fibre bituminous roofing panel sanded on the under-side.

● To protect against mechanical damage in subsequent building operations, the surface finish should be applied immediately after the completion of the sealing work.

● If the surface finish (concrete flagstones, asphalt or precast slabs laid on a mortar bed) covers the whole surface of the sealing layer, it must be separated from the surface finish by a separating layer (two layers of loose-laid polyethylene foil) in order to prevent the transfer of movement.

● Pad-support mountings with as wide a surface as possible, to balance out the unevennesses, should be used (e.g. mortar dabs); these distribute the load uniformly over a larger surface and reduce the risk of damage.

● In order to prevent the accumulation of dirt under pad-supported surface slabs, which would affect the drainage of water from the sealing layer, the individual slabs should be laid with narrow joints. Frequent cleaning, or the filling in of the voids with loose gravel, counteracts this danger.

● To provide good resistance to water penetration, balcony slabs should be constructed of precast waterproof concrete, of limited dimensions, and should be secured on cantilevers, separated from other structural members.

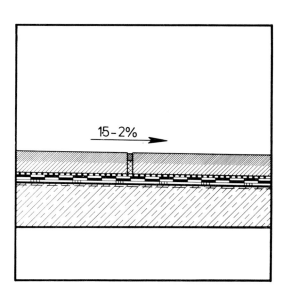

15–2%

Balconies
Sequence of layers and individual layers

In the weathering surface of balconies, loggias and open arcades which are laid in large areas with no allowance for expansion, cracks occur mainly at right angles to the front of the building, sometimes diagonally, leading from fixed points such as protruding wall corners. These cracks cause defects in the sealing layer and the reinforced concrete slab becomes saturated with resulting efflorescence and damage to surfaces, etc. (see D 1.1.2 – Linear expansion of balcony slab). The surface finish material may become loose over parts of the surface and bulge, there is sometimes failure of the edge joints and, in asphalt surfaces, edge settlement. At the junction with vertical parts of the structure, if a thin finishing coat has been applied up to the edge and pad-mounted slabs are laid, deformation and failure of the sealing layer occurs. The water penetration leads to extensive damage on the external walls and in the adjacent rooms (see D 2.1.5 – Sealing junctions made of foils or strips).

Points for consideration

— On balconies, particularly those facing south or west, the surface weathering layers are exposed to both long-term and short-term variations in temperature. Additionally there are varying dampness stresses.

— The resulting changes in length and form lead, when restricted by adjacent structural members or those that pierce the surface, to pressure stresses. The thinness and the material properties, especially the minimal tensile strength, of the surface layers result in cracks and fractures.

— Inserted reinforcement (steel mesh) improves the capacity to absorb loads (bending moments).

— The high temperature-linked linear expansion of asphalt surfaces, when restricted by vertical walls, for example, leads by compression to plastic deformation. When the temperatures fall, reducing plasticity, thermally caused contractions occur and settlement cracks appear in adjacent structural members and in the surface where, with large asphalt areas, there is defective separation from the base (sealing layer).

— Because linear expansion of the surface layers is unavoidable, transmission of stress to the vulnerable sealing layer must be prevented.

— Weathering surface finishes laid direct on to the sealing layers, in time and under temperature influence, form an internal bonding with the bituminous adhesive layers of the sealing layer.

Recommendations for the avoidance of defects

● A frictional connection between the sealing layer and the weathering surface layer must be prevented. To achieve this, a separating layer must be inserted, e.g. in the form of two layers of loose-laid polyethylene foil.

● The weathering surface must be separated by an edge joint from all adjacent structural members or those piercing it, and subdivided by expansion joints into areas with a maximum length of approx. 1·5 m (see C 2.3.3 – Expansion joints in the balcony surface).

≤ 1·50m

Balconies
Sequence of layers and individual layers

On balconies where the sealing layer has been laid level or with a defective fall, or is not connected to the drainage points, damage occurs on the weathering surface, which is saturated through cracks and expansion joints. Where there are sealed weathering layers made of stone, ceramic slabs, etc., the dampness remains for a long time. In the area of the cracks deposits of eroded mortar particles and dirt are formed, the slabs loosen and the surface area is liable to freeze.

Saturation of the reinforced concrete slab followed by erosion and soiling on the under-side occurs in balconies with no sealing layer underneath the weathering surface layer of asphalt or concrete. Rainwater penetrates the reinforced concrete slab through cracks and hairline fractures in the surface layer and in the thin coatings.

Points for consideration

– Weathering surface layers do not function as sealing membranes because of their liability to cracking.
– Dampness which has penetrated to a waterproof surface through local permeability of the weathering layer dries out slowly at the area of penetration. With frost action dampness in the waterproof surface can lead to freezing of the surface layer and to joint failures.
– Rainwater which penetrates the surface layer must be drained away as quickly as possible.

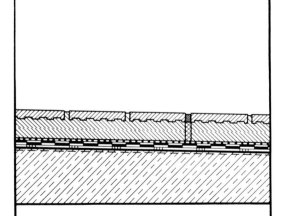

Recommendations for the avoidance of defects

● The weathering surface slabs should be applied on a sealing layer having a similar fall and on an intermediate separating layer.

● The weathering surface layer should be applied on layers which are permeable to water.

● Weathering layers of flagstones or slabs laid in a mortar bed should be set on 50 mm thick protective concrete with porous joints.

● Weathering layers made of precast slabs (stone or artificial stone) with minimum dimensions of 400 × 400 mm are laid in a level and precompressed gravel bed ⩾ 50 mm thick consisting of washed granules of 16–32 mm diameter.

● Weathering layers made of precast slabs (stone or artificial stone) can be laid on pads. In order to avoid excessive spot-loading on the roof membrane, the pad supports should cover as wide an area as possible, e.g. in the form of mortar spots.

● Balcony slabs which are freely supported in front of the façade on lateral supports (e.g. cantilevers) can be accepted as a weathering surface without an additional sealing and surface layer if they are no longer than 4 m, the reinforced concrete is waterproof and there is a trafficking finish to the upper surface.

Balconies
Sequence of layers and individual layers

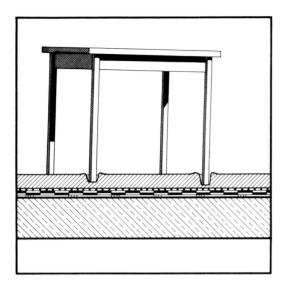

The weathering surface of the surface layer has been found to show various types of damage which seriously affect the use of balconies, loggias and arcades. Flagstones and precast slabs may be cracked, and at the edges of structural members, e.g. above plate connections, the surface layers loosen. Slabs bedded in a sand or gravel bed become uneven, with tilting of individual slabs. Applied coating films on flagstones peel off, eroded mortar particles and dirt are deposited along cracks at the edges of expansion joints in the surface. Asphalt surfaces soften, especially those exposed to sunlight, and, as a result, pressure points are formed which compress the asphalt to a thin layer under tables and chairs.

Where the weathering surface has been laid without a fall or with a defective fall, rainwater and melted snow remain as puddles, freezing in cold weather. Water loading is increased by wind banking, with resulting damp penetration at the edges of structural members and at wall connections (see D 2.1.2 and 2.2.2 – Sealing layer upstand).

Points for consideration

– If there is no fall to the weathering surface of the surface layer, if the fall is incorrect or if the catchment surface to the roof outlets is too large, rainwater drains away slowly or stands as puddles until it evaporates. In frosty weather ice forms on the upper surface.

– The surface layer must be drained as quickly as possible.

– Precast slabs bedded in gravel can be tilted by edge loading on a yielding base; or the gravel bed may be eroded when small-size granules are washed from underneath the slabs so that voids occur below them.

– The surface, especially the weathering layer, should be designed according to its specific intended use: for example, the fall should be $\leqslant 2\%$. If there are increased loads, e.g. plant containers, furniture, equipment, the surface must be designed to have appropriate loading capacity and stability.

– Under prolonged spot-loading, pressure points occur on asphalt surfaces, especially with high temperatures, and these tend to yield and shift laterally.

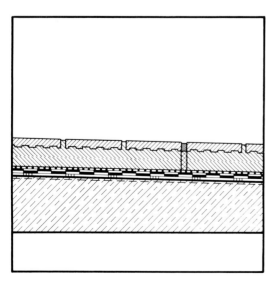

Recommendations for the avoidance of defects

● The weathering surface of the surface layer should be laid to a fall of 1·5–2% and be connected to the drainage system. Where slabs are laid in a gravel bed or on pad mountings, there is no need for a fall or a connection to the drainage of the weathering surface.

● Precast slabs bedded on a gravel fill should be $\geqslant 400 \times 400$ mm and heavy, and should be fully supported on a 50 mm thick precompressed gravel layer composed of washed granules of 16–32 mm diameter.

● Asphalt should be laid only on surfaces which are protected from direct sunlight and do not suffer any great stress from point loads.

General texts and principles

Eichler, Friedrich: Bauphysikalische Entwurfslehre, Band 2, 4. Auflage, Verlagsgesellschaft Rudolf Müller, Köln 1973.

Forschungsgemeinschaft Bauen und Wohnen, Balkone im Wohnungsbau – Erfahrungen und Beobachtungen über Anordnung und Ausführung; Bericht der FBW, Heft 35/1954, Stuttgart 1955.

Lufsky, Karl: Bauwerksabdichtungen – Bitumen und Kunststoffe in der Abdichtungstechnik, 2. Auflage, B. G. Teubner, Stuttgart 1970.

Moritz, Karl: Flachdachhandbuch – Flache und flachgeneigte Dächer, 4. Auflage, Bauverlag Wiesbaden und Berlin 1975.

AGI – Arbeitsblätter A 10 – Industrieböden, Hartbetonbeläge; A 11 – Industrieböden, Zementestrich als Nutzboden.

DIN 4122 – Abdichtung von Bauwerken gegen nicht drückendes Oberflächenwasser und Sickerwasser mit bituminösen Stoffen, Metallbändern und Kunststoffolien, Richtlinien, Juli 1968.

DIN 18338 – VOB/C Dachdeckungs- und Dachabdichtungsarbeiten, August 1974.

DIN 18353 – VOB/C Estricharbeiten, August 1974.

DIN 18354 – VOB/C Asphaltbelagsarbeiten, Februar 1961.

Supporting slab

Balkowski, Dieter: Bautechnische Kriterien der Balkonplatten. In: Das Dachdeckerhandwerk (DDH), Heft 10/75, S. 696–697.

Caemmerer, Winfried: Wärmeschutz aber richtig. Herausgeber: Deutsches Bauzentrum, Köln 1958.

Grün, Wolfgang: Balkone, Loggien, Terrassen. In: Deutsche Bauzeitschrift (DBZ), Heft 5/71, S. 937–950.

Zimmermann, Günter: Plattenbeläge auf Balkonen. In: Architekt + Ingenieur, Heft 5/71, S. B1–B4.

Sealing layer

Gäbges, Peter: Die Fragwürdigkeit von Plattenkunststofflagern auf abgedichteten Terrassen. In: Deutsches Architektenblatt (DAB), Heft 11/71, S. 404–405.

Probst, Raimund: Außengänge, Balkone, Dachterrassen. In: Baupraxis, Heft 3/70, S. 39–43.

Schütze, Wilhelm: Der Estrich auf Dächern und Terrassen, eine Estrichart, die ganz besondere Sorgfalt erfordert. In: Boden, Wand + Decke, Heft 3/65, S. 182–192; Heft 4/65, S. 294–304.

Zimmermann, Günter: Dachterrassen mit aufgestelzten Belägen, Durchfeuchtungen von Decken und Wänden. In: Deutsches Architektenblatt (DAB), Heft 12/74, S. 927.

Zimmermann, Günter: Plattenbeläge auf Balkonen. In: Architekt und Ingenieur, Heft 5/71, S. B1–B4.

Österreichisches Institut für Bauforschung: Abdichtungen und Abläufe bei Flachdächern, Dachterrassen, Balkonen, Loggien, Naßräumen; Forschungsbericht 39, 2. Auflage, Wien 1973.

Surface layers

Bech, Helmut: Stellungnahme zum Beitrag »Plattenbeläge auf Balkonen«. In: Architekt + Ingenieur, Heft 7/71, S. 35–36.

Gäbges, Peter: Terrassen, Plattenlager. In: Das Dachdeckerhandwerk (DDH), Heft 21/72, S. 1553–1554.

N. N. – Fliesen als Balkon- und Terrassenbeläge. In: Deutsches Architektenblatt (DAB), Heft 14/73, S. 1145–1146.

Rick, Anton W.: Risse und Randabsetzungen bei Asphaltterrassenbelägen. In: Bitumen, Teere, Asphalte, Peche... (BTAP), Heft 10/73, S. 422–428.

Rick, Anton W.: Einiges über Asphaltbeläge auf Terrassen. In: Bitumen, Teere, Asphalte, Peche... (BTAP), Heft 2/72, S. 82–83.

Schütze, Wilhelm: Der Estrich auf Dächern und Terrassen, eine Estrichart, die ganz besondere Sorgfalt erfordert. In: Boden, Wand + Decke, Heft 3/65, S. 182–192, Heft 4/65, S. 294–304.

Zentralverband des Dachdeckerhandwerks: Richtlinien für die Ausführung von Flachdächern, Ausgabe Januar 1973, Helmut Gros, Fachverlag, Berlin 1973.

Zimmermann, Günter: Plattenbeläge auf Balkonen. In: Architekt + Ingenieur, Heft 5/71, S. B1–B4.

Zimmermann, Günter: Balkon aus Stahlbetonplatte und Spaltklinkerbelag im Mörtelbett. Rissebildung im Belag und in Anschlußfugen. In: Deutsches Architektenblatt (DAB), Heft 10/73, S. 857–858.

Problem: Connection to vertical abutments

With balconies, loggias and open arcades continuity of construction of the structural members with that of the vertical external walls must be ensured. In some cases a connection to parapet units, plant containers, etc., is also necessary. Depending on the design of the balcony slab, as a cantilever slab or supported on lateral bearing in front of the façade, an appropriate structural solution for the connection must be selected. This must prevent overstressing of the vertical wall by damp penetration from the balcony surface and must ensure the correct functioning of the balcony sealing layer.

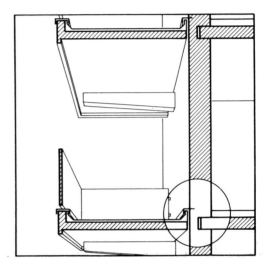

Damaged structural members have been found without exception to be designed as cantilevered slabs. The connection to the structural walls provides the most difficult problem with balconies, loggias, arcades, and this is where most of the defects occur.

A few, frequently recurring defects in detail design and construction are accountable for the majority of failures. On the following pages the defects are analysed with suggested recommendations for their avoidance. The detail solution for balcony slabs with lateral support is also investigated in depth.

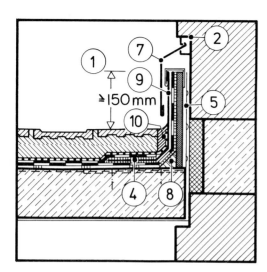

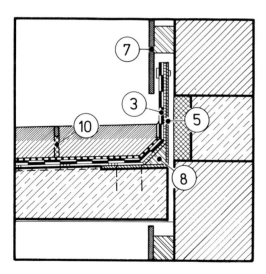

1 At the angle of the balcony with the vertical structural wall, the sealing layer must be raised above the highest level of water that may be expected. The height of the upstand should be 150 mm above the weathering surface of the surface layer (see D 2.1.2).

2 The required structural height of the edge must be established at design stage. This is calculated from the height of the upstand and the number and thickness of built-up layers (see D 2.1.2).

3 Sealing angles using wall connection plates should be avoided. The junction of the sealing layer to the vertical wall surface should be formed from the sealing layer itself or by use of connection foils of similar material properties (see D 2.1.4).

4 If the use of metal connection plates is unavoidable, they should be protected from the possible chemical action of fresh lime and cement mortar by bitumen coating. Expansion joints at adequate centres must be provided to allow for movement. After the removal of damp and dirt, the metal plate should be bonded in ≥ 120 mm between the individual layers of the multi-layer roof membrane (see D 2.1.4).

5 If the balcony is constructed as a slab with lateral support then the sealing layer should be raised direct on the balcony slab and fixed independently of the façade. The water-proofing of the connection joint can be achieved with a metal cover flashing (see D 2.1.3).

6 If, on a balcony slab with lateral support, the angled sealing layer has to be fixed to the façade, then to accommodate the various movements, it is necessary to use overlapping but freely moving pressed metal sections (see D 2.1.3).

7 At the junction with the vertical surface, the edge of the sealing layer must be protected from the water streaming down the wall face. This can be achieved by recessing the edge of the sealing layer behind the water-diverting surface of the wall, or by protecting it with a cladding (see D 2.1.3).

8 At the junction the sealing layer should be raised up, without sharp bends, over a preformed edge fillet to ensure a fully supported sealing base (see D 2.1.5).

9 The angled sealing layer should be protected by the formation of an upstand or by suspending a metal cover flashing in front of it (see D 2.1.3).

10 The balcony weathering surface layer must be separated from the angle upstand by an edge joint taken to the separating layer (see D 2.1.5).

Balconies
Connection to vertical abutments

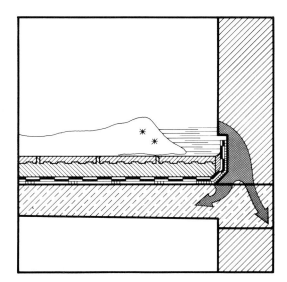

Damp penetration of reinforced concrete cantilever slabs is frequently observed. This results in failure of the roof surface and loosening of flagstones, efflorescence, soiling and spalling of the rendering on the under-side of the balcony slab. In all such cases the sealing layer has been laid flat or reaches only to the height of the weathering surface layer, and is at that point finished with mastic, or the angle is only a few centimetres above the surface.

If there is no sealing layer under the surface layer, rainwater penetrates the edge area of the surface and the cracks in the weathering layer.

Points for consideration

– The vertical structural members – particularly walls – in the area of the adjacent balcony surface are exposed to strong stresses through water spray and banked-up water, such as normally occur only in the foundation area of buildings. Special measures for protection against water are necessary here.

– Where there is no fall or only a shallow fall, there can be intense water penetration. To avoid this additional stress, the surface and sealing layer should therefore be laid to falls away from the vertical structural member.

– In practice, because of the unevennesses, the junction of the angle of the raised sealing layer to the vertical surface of the wall is difficult to waterproof. It must be protected from water running down the wall surface.

– Where balcony slabs are supported on lateral bearings (e.g. cantilevers) which are built into the wall, damage to the wall surface is minimal if the slope of the balcony slab itself is adequate.

– Between the façade and the separated fixed balcony slab movements, normally parallel to the façade, occur as a result of temperature influence.

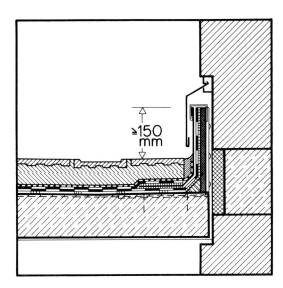

Recommendations for the avoidance of defects

● In the abutment area the sealing layer must be raised above the highest level of water that may be expected. The height of the upstand should be 150 mm above the weathering surface of the surface layer.

● Where balconies are constructed as slabs on cantilevered bearings the sealing layer should be formed into an upstand on the balcony slab itself, independent of the façade. The joint between the angled sealing edge and the façade can be protected by metal cover flashings.

● The required structural height of the angle, calculated from the upstand height and the number and thickness of surface layers and falls, must be established at design stage.

Balconies
Connection to vertical abutments

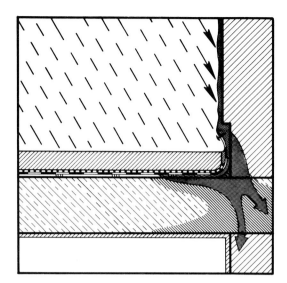

Rainwater may penetrate behind the edge of the raised sealing layer at the connection to the vertical wall surface. This connection leaks when the sealing layer is fixed to the upright surface only by adhesion and comes loose in places or along its whole length. Cover flashings made of sheet metal, inserted into joints, may slip out.

Points for consideration

— The higher a vertical surface and the more freely it is exposed to the weather, the heavier is the stream of water that runs down its surface.

— If the junction of the sealing layer is placed in front of the façade, it obstructs the stream of water and is more strongly stressed.

— The upstand sealing layer is subject, at the change of plane, to strong thermal influences which reduce its stability and that of the adhesive layers.

— The surface of the wall is usually uneven so that a metal cover flashing has no smooth bearing surface. Sealing by means of cement has limited durability.

— In examples of balcony slabs on lateral bearings, e.g. cantilevers, strong movements, mainly parallel to the connection joint, can be expected between the edge of slab and the façade, due to the temperature stress on the balcony slab.

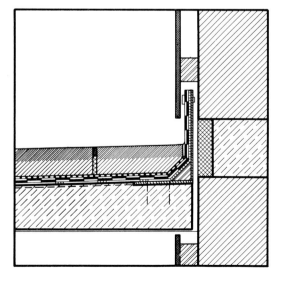

Recommendations for the avoidance of defects

● The edge of the sealing layer must be protected from water flowing down over its junction with the wall surface. This can be achieved either by recessing the sealing edge behind the water-diverting wall surface or by inserting it behind a cladding layer.

● If the balcony is constructed as a slab with lateral bearings the sealing layer must be raised on the balcony slab itself and fixed independently of the façade. The sealing of the junction can be achieved with a cover flashing.

● If, on a balcony slab with lateral support, the angled sealing layer has to be secured to the façade, then to accommodate the various movements, it is necessary to use overlapping but freely moving pressed metal sections.

● The angled sealing layer should be protected by the formation of an upstand or by suspending a metal cover flashing in front of it.

Balconies
Connection to vertical abutments

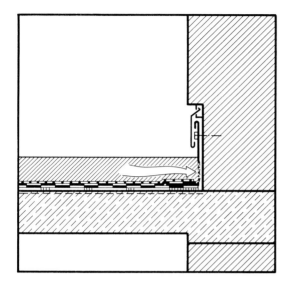

If pressed metal or metal sections are used in the construction of the angle of the sealing layer to vertical surfaces (walls, parapets), then these metal parts frequently let in water. Metal sections in long lengths without provision for expansion show cracked soldered seams, considerable bulging and warping. In some cases the bonding of the section with the sealing layer is not watertight, because the sections have been bonded only on one side to the sealing layer. Lead or zinc sections laid in a mortar bed without protection are subjected to corrosive attack and become porous.

Points for consideration

– Angles of the sealing layer which are not protected – by raised slabs, for example – are subject to climatic influences and are particularly exposed to changes in temperature.

– The effect of linear expansion in sections of longer lengths can be reduced by subdividing into shorter sections and by the installation of overlapping sleeves. It should be understood that the construction of these expansion joints is expensive in terms of labour and they conceal new weaknesses; if possible, sections bonded firmly with the sealing layer should not be used.

– The use of pressed metal angles necessitates the permanent bonding of two very different structural materials by means of an adhesive.

– Metals – above all milled lead and zinc – are chemically vulnerable to fresh lime or cement mortar and concrete: the calcium hydroxide, $Ca(OH)_2$, dissolves the metal. A direct contact must therefore be avoided.

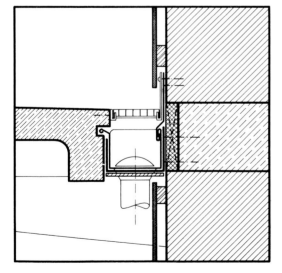

Recommendations for the avoidance of defects

● Sealing angles made from metal sections should be avoided. The angle of the sealing layer to the vertical surface should be formed from the roof membrane itself.

● If the use of metal sections is unavoidable, they should be protected with a coating (e.g. bitumen) from the chemical action of fresh lime and cement mortar. They should be provided with overlapping sleeves for expansion at adequate centres (with titanium zinc, $\geq$ every 3 m). After the removal of damp and dirt, the sections should be inserted and bonded in at least 120 mm between the individual layers of the multilayer roof membrane.

● The angled sealing layer should be protected by the raising of the surface finish or by sheet metal cladding suspended in front of it.

Balconies
Connection to vertical abutments

Where angles are formed out of the sealing layer itself or with the use of foils – e.g. from synthetic material – various defects are found to occur as the result of mechanical stresses. Where the change of plane deflects the sealing layer in a right angle, at the junction where it is unsupported the sealing layer may be pierced. In other cases, the surface finish is taken up to the angle without a separating joint, and the sealing layer is damaged by movement in the surface finish.

Points for consideration

– During the construction of the surface finishing layer, the unprotected sealing layer undergoes strong mechanical stresses from foot traffic, passage of wheelbarrows, storing of construction materials, etc.

– Multi-layer sealing layers and foils cannot be deflected into a right angle. At sharp angles the bituminous bonding layers crack and the sealing layer is unsupported.

– The vulnerability of the sealing layer to mechanical damage is increased in places where it is unsupported.

– After completion, the surface finishing layer is exposed to considerable changes in temperature and is subjected to corresponding thermal linear expansion. As these surface layers are normally loosely applied to the sealing layer, the linear expansion occurs at the edge of the interacting surfacing, e.g. at the wall connection, where the vulnerable sealing layer is angled.

Recommendations for the avoidance of defects

● The angle of the sealing layer to surfaces of vertical structural members should be formed from the roof membrane itself or by use of connection foils of similar material properties.

● At the junction the sealing layer should be raised without sharp deflections over an angle preformed in the sealing base, ensuring a continuous sealing support.

● The balcony surface must be separated from the angle by an edge joint taken to the separating layer (see D 1.1.5 and 2.3.3 – Expansion joints in the balcony surface).

Problem: Edge of structural member

As with flat roofs and roof terraces, the edge of the balcony must be designed on the principle that all structural members must be waterproofed with controlled drainage of water from their surfaces. Rainwater overflowing the edge of the structural member is acceptable only in secondary balconies overhanging green landscaped areas. In principle, all structural members drained outwards must be terminated with pre-set gutters; all structural members drained inwards must have an angled upstand which raises up the sealing layer.

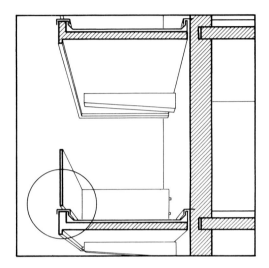

Just as the problem of sealing of the balcony surface is frequently neglected (see D 1.1.4 – Sealing layer), in most cases the edge indicates an inadequate structural formation.

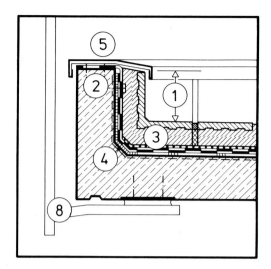

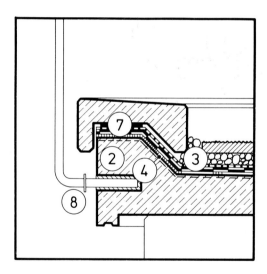

1 The sealing layer must be taken to the edge of the structural member and raised to a height of $\geqslant 100$ mm (see D 2.2.3).

2 At the edge of inwardly drained balconies, loggias and open arcades the sealing layer must be raised $\geqslant 100$ mm above the level of the surface finish (see D 2.2.2).

3 The necessary structural height of the edge must be established at design stage. This is calculated from the height of the upstand and the number and thickness of built-up layers (see D 2.2.2).

4 At upstand edges the sealing layer itself should be raised on a slope preformed in the reinforced concrete slab (see D 2.2.4).

5 A frictional connection of the sealing layer with metal edge sections should be avoided (see D 2.2.4).

6 The sealing layer may only be connected to parapet units if these are permanently sealed against rainwater. The sealing layer must be raised to a height of 150 mm and be secured with a cover flashing to the sealing edge (see D 2.2.3).

7 If the parapet units at the edge of the structural member (e.g. plant containers) are not permanently watertight or have joints which are difficult to seal, the sealing layer must go under those units and to the edge of the structural member (see D 2.2.3).

8 Balustrade fixings on the upper side of the balcony slab, which pierce the surface finish and sealing layer, should be avoided if possible. Balustrades should be fixed outside the sealed surface of the structural member in the loadbearing parts, at the front edge or under-side of the balcony slab. In small balconies, a fixing on the lateral wall faces is possible (see D 2.2.5).

9 If fixing to the upper surface of the structural member cannot be avoided, the balustrade should be set into an edge upstand, which is fixed to the balcony slab. Balustrade posts which pierce the sealing layer should be inserted in the reinforced concrete slab. The connection to the sealing layer must be achieved independently of the post by means of a socket having a fixed and loose flange; this is sealed to the balustrade post by means of a cover cap. The surface finish must be separated by an edge joint (see D 2.2.5).

Balconies
Edge of structural member

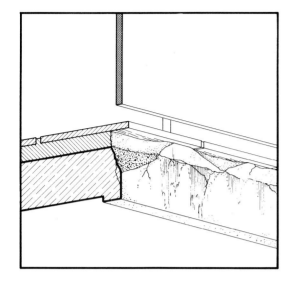

A number of balconies, loggias and open arcades show no design consideration for the termination of the structural member at the edge: the sealing layer and the surface finish end at the edge of the cantilevered slab. In some examples, edge trims are installed, designed to prevent dripping. In most of these structural members without edge detailing, the result is water cascading from the upper surface, and cracking and spalling of the render layer on the edge. These defects are worse in structural members which have a defective fall or a fall towards the edge in the sealing layer or in the surface finish.

Points for consideration

— Rainwater is drained to various parts, according to the surface finish selected, on the upper surface and on the sealing layer. Both layers must therefore be laid to falls towards the outlet points.

— The formation of puddles, backwash and wind banking can result in water penetration and overflow at the edge section if there is no adequate upstand to the sealing layer.

— Where the edge of a balcony has an upstand, rainwater or domestic water can bank up if outlets are blocked. The height of the angle of the sealing layer at the junction with vertical walls and thresholds must be above the highest expected water level.

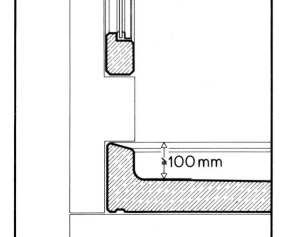

Recommendations for the avoidance of defects

● At the edge of inwardly drained balconies, loggias and open arcades the sealing layer must be raised ⩾ 100 mm above the level of the surface finish.

● The required structural height of the edge must be established at design stage. This is calculated from the height of the upstand and the sequence of surfaces and falls.

Balconies
Edge of structural member

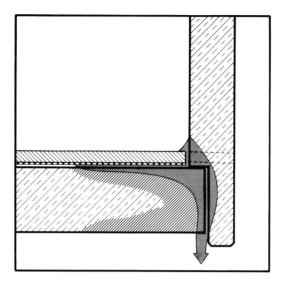

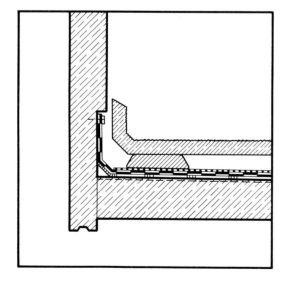

In the transition area of the balcony surface to solid parapets or plant containers, leakages are found which cause saturation damage to the under-side of slabs, with dripping, soiling and spalling of finishing layers. In some cases water runs under the sealing layer and leads to damp penetration to the floors inside.

Where plant containers are placed on the edge of the structural member and the sealing layer is connected to them, damp penetration may also occur. The rainwater penetrates through the plant tubs and the open expansion joints into the surface of the reinforced concrete slab, and drips down the front edge of the balcony.

Points for consideration

– On the upper surface of the structural member of parapets, etc., a permanently watertight connection of the angled sealing layer by means of bonding, clamp rails or cement is not possible. This junction must be protected from the vertical stream of water.

– Plant containers are not usually completely watertight in the long term; rainwater enters through the sides or the floor of the container into the bearing joint.

– Precast elements such as plant containers and parapets are provided with open joints. These expansion joints are difficult to seal against water.

Recommendations for the avoidance of defects

● The sealing layer must be taken to the edge of the structural member and be raised to the required height, ⩾ 100 mm.

● If the parapet elements at the edge of the structural member (e.g. plant containers) are not permanently watertight or have joints which are difficult to seal, the sealing layer must go under them and up to the edge of the structural member.

● The sealing layer may be connected to parapet elements only if these are waterproof. The sealing layer must be raised to an angle height of 150 mm and be fixed to the sealing edge with a cover flashing (see D 2.1.2 – Sealing layer upstand).

Balconies
Edge of structural member

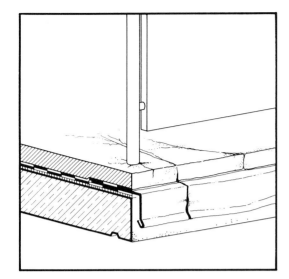

At the eaves and edges of structural members, if metal sections or sheets have been bonded into the multi-layer sealing, the result is warping and bulging of the sheets, opening of the expansion joints and splitting of the soldered seams with tears in the sealing layer. The rainwater saturates the cantilevered slab and causes soiling and discoloration on its under-side. In the edge area of structural members where the sealing layers are laid level, the rainwater does not completely drain off the structural member but gradually penetrates the surface finish. If the surface finish has been laid over bonded-in metal sheets, the surface layers come loose from the metal sheets with total weathering failure in the edge area.

Points for consideration

– The termination of the sealing layer at the balcony edge by means of metal sections requires the permanent direct bonding of two very different materials with an adhesive.

– Edge sections are exposed without protection to extremes of climate. In metal sections there is a considerable change in length and form, which can excessively stress other adjoining structural elements which are frictionally connected with them. This can lead to failure by compression (warping) and tension (cracks).

– Provision for expansion in the metal sections (e.g. sliding sleeves) to reduce effective expansion lengths are expensive in labour and conceal new weaknesses.

– Edge trims on balconies with a shallow, or no, fall to the sealing layer may, as a result of temperature-induced deformations, produce a counterfall towards the surface of the structural member. This prevents the complete drainage of the surface and frost action can lead to freezing and further damage to the edge.

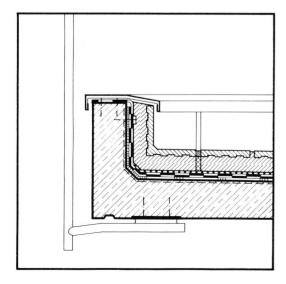

Recommendations for the avoidance of defects

● A frictional connection of the sealing layer with metal edge sections should be avoided.

● Where there are angled edges the sealing layer itself should be raised on a preformed fillet in the reinforced concrete slab.

165

Balconies
Edge of structural member

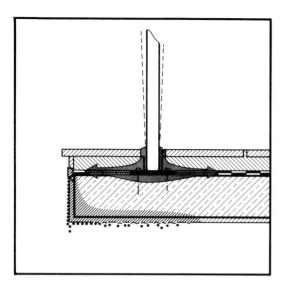

At edges of balconies with balustrades, cracks and spalling rendering appear at the outer edges, resulting in saturation of the cantilevered slab with soiling and discoloration on the under-side. This defect occurs mainly where balustrade posts are fixed from above in the reinforced concrete slab. The pierced sealing layer may be quite unfixed, or only a cement fillet provided. In some balconies the balustrade posts are very slender and are secured to the balcony slab only by means of a small base plate, so that flexible movements cause further damage.

Where railing posts have been inset into half-height reinforced concrete parapets, the concrete rendering cracks and spalls with consequent corrosion of the steel reinforcement.

Points for consideration

— Balustrade posts fixed in the reinforced concrete slab from above pierce the surface finish and sealing layer and form a vulnerable point for subsequent failure.

— The balustrade posts, through usage and wind, are horizontally loaded. The resulting flexible movements and bending stress the sealing layer and surface finish.

— Where balustrade posts are fixed directly to the edge upstand or slab, rainwater collects in the pockets that are unavoidably formed between the posts and the grouting; this leads to frost damage and spalling of the concrete and to corrosion of the posts.

Recommendations for the avoidance of defects

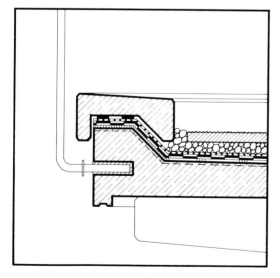

● Balustrade fixings on the upper side of the balcony slab which pierce the surface finish and sealing layer should be avoided. The balustrade should be fixed outside the sealed surface of the structural member in the loadbearing parts in the front edge or under-side of the balcony slab. With small balconies, fixing to the lateral wall faces is possible.

● If fixing on the upper surface of the structural member cannot be avoided then the balustrade should be set into the edge upstand, which is fixed to the balcony slab. Balustrade posts which pierce the sealing layer should be set in the reinforced concrete slab. The junction with the sealing layer must be achieved independently of the post by means of a socket with a fixed and loose flange; this is sealed to the balustrade post by means of a cover cap. The surface finish must be separated by an edge joint.

Problem: Expansion joints

Balconies, loggias and arcades as horizontal structural members are exposed on all sides to the external climate and are also exposed to strong variations in temperature, but cannot, in contrast to roofs and roof terraces, be protected from these stresses by thermal insulation.

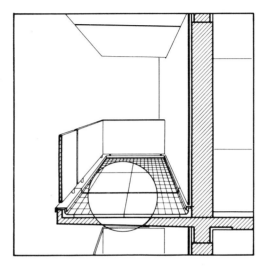

The expected linear expansion must be accommodated by structural measures. The arrangement of expansion joints is important, especially in reinforced concrete cantilevered slabs connected directly with the adjoining floor slab.

On the following pages structural recommendations for this type of problem are illustrated. Surface joints, of which the problems are largely the same as for those in roof terrace surfaces, are also dealt with.

Balconies

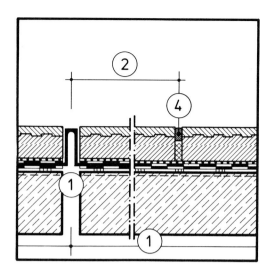

1 Balcony slabs and concrete parapets should have expansion joints at maximum centres of approx. 4 m. The expansion joint should be carried through the structural member with a width ≥ 20 mm. On the sealing layer the expansion joint must be covered by bonded-in foils allowing for movement, or expanded metal sections (see D 2.3.2).

2 Balcony surfaces must be subdivided by expansion joints into areas with a maximum length of approx. 1·5 m and must be separated by an edge joint from all adjacent parts of the structure (see D 2.3.3).

3 Where slabs are loose laid in gravel or on pad supports, there must be open joints between the individual slabs; joint width depends on the slab dimensions, but must be ≥ approx. 5 mm (see D 2.3.3).

4 Where flagstones or other surface finishes are laid on a mortar bed, the surface expansion joints should be continued for a width of ≥ 10 mm through the complete balcony surface structure as far as the separating layer. If these joints cannot be left open, the joint in the lower section should be filled with a non-rotting, flexible material (e.g. strips of polystyrol foam) and covered on the upper surface by a gasket or joint compound (see D 2.3.3).

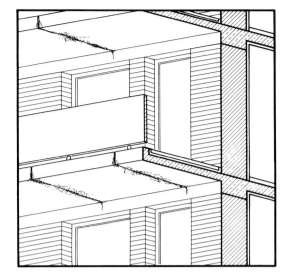

When reinforced cantilevered slabs with a length > 4 m are fixed on one side only and are not subdivided by expansion joints, they frequently show cracks at right angles to the elevation. This failure usually continues through the sealing layer and the surface finish and results in damp penetration, erosion and efflorescence on the under-side of the slabs.

Where expansion joints are provided only between firmly fixed precast parapet slabs, these joints continue as cracks in the balcony cantilevered slab.

Points for consideration

— The balcony slab is exposed to strong variations in temperature and dampness and subject to considerable linear expansion which is restricted, particularly where the cantilevered slab is fixed on one side only. The resulting compressive and tensile stresses can lead to damage by cracking.

— Structural members which are exposed to the external climate on all sides cannot be protected by thermal insulation; the linear expansion must be contained by the structure to avoid damage. One method is to subdivide the balcony slab by expansion joints (see also D 1.1.2 – Linear expansion of balcony slab).

— If expansion joints are not continued through the whole balcony and the surface layer then they continue as cracks in the areas not subdivided.

Recommendations for the avoidance of defects

● Balcony slabs and continuous concrete parapets should have expansion joints at approximately 4 m centres. The expansion joint should be continued through all layers of the structural member at a width of ⩾ 20 mm. In the sealing layer the expansion joints must be covered by bonded-in foils allowing for movement, or by expanded metal sections.

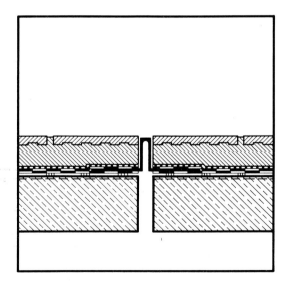

Balconies
Expansion joints

If balcony surfaces of concrete flagstones or slabs are laid in a mortar bed over several metres and are not subdivided by expansion joints, then cracks, warping, unevennesses and efflorescence appear on the surfaces.

Points for consideration

– Surface layers are subjected to large variations in temperature and therefore undergo considerable linear expansion. If movement is restricted, pressure stresses occur which can be absorbed only by a multi-layer form of construction, and to a limited extent this avoids cracking or warping.

– If the sequence of layers of the balcony surface is carried out correctly, in that a separating layer is inserted which allows the surface finish to be loosely connected with the sealing layer, and the lower surfacing layer can drain away any seepage water without subsequent frost damage (see D 1.1.4 – Sealing layer), the linear expansion can be absorbed by an arrangement of expansion joints in the surface layers.

– Expansion joints in the surface layers do not need to be sealed. If joint fillers are necessary, then filling with a flexible material should prevent the joint from obstruction and thus, in certain circumstances, from losing its capacity to move freely.

Recommendations for the avoidance of defects

● Balcony surfaces must be subdivided by expansion joints into areas that have a maximum length of approximately 1·5 m and must be separated by an edge joint from all adjacent vertical structural members or those which pierce the surface.

● Where slabs are loosely laid in gravel or on pad supports there must be open joints between the individual slabs. The joint width is dependent on the slab dimensions, but should be ⩾ about 5 mm.

● Where flagstones and other surface finishes are laid on a mortar bed, the surface expansion joints should be continued in a width of ⩾ 10 mm through the whole balcony surface structure as far as the separating layer. If these joints cannot be left open, the joint in the lower area should be filled by non-rotting, flexible material (e.g. strips of polystyrol foam) and should be covered on the upper surface by a gasket or joint compound.

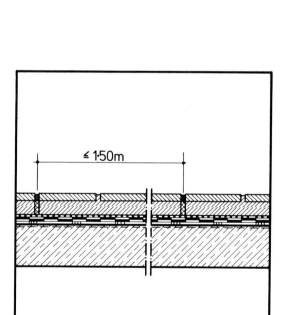

≤ 1·50 m

Problem: Drainage

If balconies, loggias and arcades are to be kept free of damage, care has to be taken to provide an adequate drainage system, although they are relatively small and only a little rain falls on them. Sealing layer and surface finish layers must be connected by means of falls to the drainage outlets in such a way that the rainwater is drained off – surface layers alone cannot undertake the task of drainage.

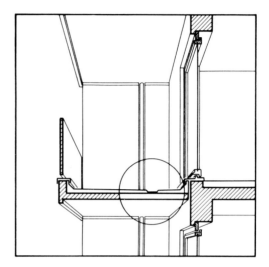

Defective design and construction result in damp penetration of the loadbearing slab as well as of the surface layers, leading in turn to efflorescence and damage from frost action. Often, no thought is given to drainage capacity: no fall is provided, or the sealing layers are imperfectly bonded to the flanges of the drainage outlet so that water penetrates into the structure beneath.

Balconies

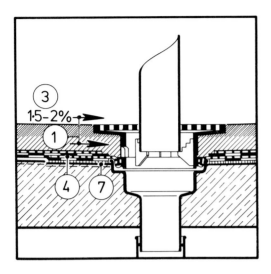

1 On balconies, loggias and arcades rainwater must, by means of suitable drainage outlets, be drained from the upper surface as well as from the sealing layer (see D 2.4.2).

2 External gutters should be used only on shallow balconies (see D 2.4.2).

3 Where balconies are drained inwardly, the rainwater must be ducted away by means of an effective fall (1·5–2%) to outlets situated in the lowest part of the roof (see D 2.4.2).

4 The sealing layer must be connected to the drainage elements with the aid of a sufficiently large flexible connection foil or with a two-part fixed and loose flange (see D 2.4.3).

5 Because of the difficulty of their connection with the sealing layer, gargoyles should be avoided as the main form of drainage (see D 2.4.3).

6 External gutters should be ≤ 3 m long. The fascia trim should be well bonded into the sealing layer (see D 2.4.3).

7 The increased thickness of the sealing layer, caused by bonding with the flange plate of internal outlets or with the fascia trims of external gutters, should be compensated by a corresponding recess in the loadbearing slab (see D 2.4.3).

8 Where slabs are laid on loose gravel, washed granules with a coarseness of ≥ 16 mm dia should be used (see D 2.4.3).

Balconies
Drainage

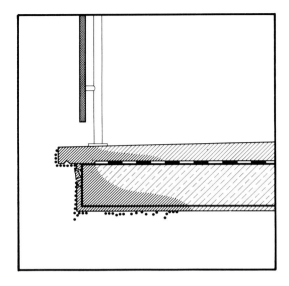

Design solutions for the drainage of rainwater from balconies, loggias and arcades are often defective in that water runs over the edge on to the balconies below. If the fascia of the load-bearing slab is terminated with precast slabs, or rendered, there is efflorescence and spalling of the finish.

Where the fall to the drainage outlets is inadequate, damp penetration appears on the lower surface together with frost action to the surface layer as a result of ponding. The same defects occur when the outlet elements are laid proud of the balcony surface.

Points for consideration

— Where balconies have no drainage outlets (external gutters or internal outlets) there is damp penetration at the edge, the water runs over the fascia and underneath the slab surface and sealing layer.

— Rainwater penetrates to the back of the rendering or slab claddings from the sides of the fascia; efflorescence and spalling result.

— External gutters require a deep roof structure to accommodate their necessary falls. With deep balconies, unusually large slopes or steps may be required.

— Individual outlets and gutters can effectively drain rainwater only from that part of the balcony surface where the fall is directed to them. The formation of puddles on the surface layer or on the sealing layer should be avoided. They lead to erosion and/or saturation of the loadbearing slab where there are even small defects in the sealing layer. Direction of fall, construction of the drainage outlet and possible sagging of the loadbearing slab should all be taken into consideration.

Recommendations for the avoidance of defects

● It should be possible to drain balconies, loggias and arcades from their upper surfaces as well as from the sealing layers by means of suitable drainage outlets.

● External gutters should be used only on balconies of small depth.

● Where balconies are drained internally, the rainwater must be ducted by means of an effective fall (1·5–2%) to outlets positioned in the lowest parts of the balcony slab.

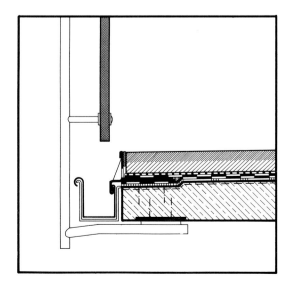

Balconies
Drainage

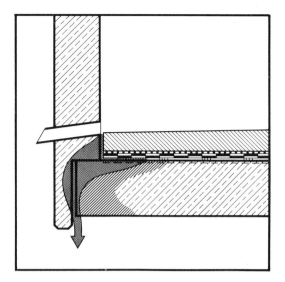

With internal drainage outlets as well as with external gutters, and especially with gargoyles, connections to the surfacing layers to be drained may be incorrectly made. The function of the sealing layer is sometimes underestimated, and the sealing layer is improperly connected to the drainage outlet or not connected at all. Water penetrates through the upper surface finish on to the sealing layer and cannot readily flow away but lies in puddles under the surface finish. Damage occurs with the freezing of the surface slab and damp penetrating the structure below.

Points for consideration

— If the loadbearing slab is not made of waterproof reinforced concrete, the upper surface layer as well as the sealing layer should be properly connected to the drainage system.

— Gargoyles should drain rainfall from both the surface and the sealing layer. An effective connection to the sealing layer is difficult to achieve because of their small dimensions. Excess loading by flowing water (including cleaning water) in deep-lying balconies must be avoided.

— If external gutters are used over long lengths ($\geqslant$ about 3 m), overlapping sections are necessary in the gutters. A waterproof connection with the sealing layer is then no longer assured.

— The connection of the sealing layers with the outlet flanges, plates or foils leads to a local increase in thickness of the sealing layer, which prevents the rainwater from completely flowing away.

— If slabs are laid in loose gravel the granules must be of sufficient coarseness not to be washed towards the outlets and block them.

Recommendations for the avoidance of defects

● Even in balconies, loggias and arcades, the sealing layer should be connected to the drainage outlets with a fall of 1·5–2·0% by means of a sufficiently large flexible connection foil or with the aid of built-up fixed and loose flanges.

● Because of the difficulty of their connection with the sealing layer, gargoyles should be avoided if possible as the main drainage outlets.

● External gutters should be $\leqslant 3$ m long. The fascia trim should be carefully bonded into the sealing layer.

● The increase in thickness of the sealing layer caused by bonding with the flange plates at the outlets or with the fascia trims of external gutters should be compensated in the loadbearing slab.

● Where slabs are laid on loose gravel, washed granules of a diameter $\geqslant 16$ mm should be used.

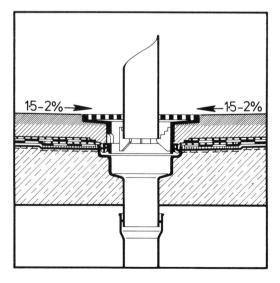

1·5–2% →   ← 1·5–2%

Problem: Thresholds

In the construction of balconies, loggias, open arcades – or any usable horizontal external surface of a building which is open on all sides to the weather – a waterproof construction of the door threshold between internal floors and balcony surface is required.

The same construction principles apply here as for the connection to vertical structural members. However, there are problems in the connection to window or door frames and the height requirement of a threshold or step between internal rooms and balcony surfaces.

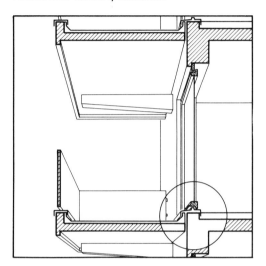

Investigation into structural defects has shown that this transition area at the door threshold poses a problem in balconies, loggias and open arcades. Numerous examples of defects of structural members have been observed at door thresholds. The analysis of defects shows recurring faults in design and construction which allow damp penetration to internal floors or in adjacent rooms.

Balconies

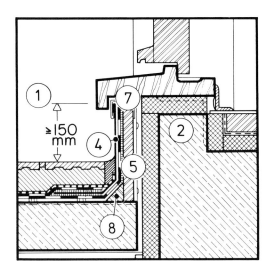

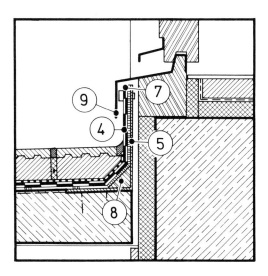

1 At the threshold of doors and windows the sealing layer must be raised above the highest expected level of water. The end of the sealing layer must be raised ≥ 150 mm above the upper surface of the balcony (see D 2.5.2).

2 The required structural height of the threshold is established at design stage, with the top edge levels of internal floors and balcony slab, and is calculated from the necessary upstand angle height and the sequence of surfacing and fall layers (see D 2.5.2).

3 If the required threshold upstand is not possible for reasons of traffic (wheelchairs, etc.) the transition from an internal room to the balcony surface must be protected from the weather by recessing the door as far back from the outer elevation as possible and by providing a steeper fall to the balcony surface (see D 2.5.2).

4 The construction of the door threshold using pressed metal sections should be avoided in thresholds > 3 m long. The angle and the junction of the sealing layer to the threshold should be formed from the roof membrane itself or by sections of material of similar characteristics (see D 2.5.4).

5 If the balcony is constructed as a slab with lateral support, the sealing layer should be raised on the balcony slab itself and be fixed independently of the façade. The waterproofing of the joint can be achieved by means of cover flashings (see D 2.5.3).

6 If the angled sealing layer has to be secured to the façade on a balcony slab with lateral supports, the junction should be made with flexible synthetic foil.

7 The edge of the sealing layer at the threshold must be protected from water flowing down the door and window surfaces. This is achieved by placing it under the threshold section (pressed metal trim) or the window ledge (see D 2.5.3).

8 At the edge the sealing layer should be smoothly raised, with no sharp deflections, over a fillet preformed in the sealing base, ensuring a complete sealing surface (see D 2.5.5).

9 The angled sealing layer must be protected by the raising of the surface slabs or by freely moving sheet metal claddings suspended in front of it (see D 2.5.5).

Balconies
Thresholds

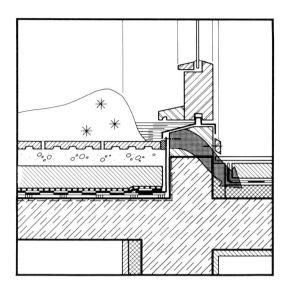

The majority of serious defects in thresholds to external doors (e.g. balcony doors) occur where there is inadequate, or non-existent, upstand angle of the sealing layer at the door threshold junction. In many cases, although the threshold is raised above internal floor level, provision of an adequate angle of fall outside means that insufficient upstand height is available, since the cantilever slab is effectively on one continuous level. Where structural members with asphalt surfaces and self-sealing flagstones are laid without an additional sealing layer, damage results from inadequate falls to the surface layers and settlement cracks form near the edge.

Driving rain, wind-banking and melted snow may cause water to overflow the sealing end or the threshold, causing damage to the floor of the room internally. In some cases the loadbearing slab also becomes saturated and efflorescence and dripping appear on the under-side of the slab.

Points for consideration

– As in other junctions of horizontal building surfaces to vertical structural members, the threshold junction is, in certain circumstances, subjected to extraordinary stresses from melted snow, banked-up water and spray, such as normally occur only in the foundations of buildings.

– The structural depth of the balcony construction, because of the required falls, is usually greater than the structural depth of the internal floors (e.g. loosely laid flagstones).

– The required height difference between the balcony surface and the threshold can be achieved by lowering the cantilevered slab in relation to the internal floor. If the balcony slab is constructed independently of the façade on cantilevers, the required height differential in relation to the threshold can be achieved without difficulty.

Recommendations for the avoidance of defects

● At the threshold of doors and windows the sealing layer must be raised above the highest water level anticipated in the worst climatic conditions. The sealing layer end must extend ≥ 150 mm above the upper surface of the roof.

● The required structural height of the threshold is established at design stage, with the top edge levels of internal floors and balcony slab, and is calculated from the necessary upstand angle height and the sequence of surfacing and fall layers.

● If the required threshold upstand cannot be constructed for reasons of traffic (e.g. wheelchairs), the transition from the inside to the balcony surface must be protected from the weather. This can be achieved by recessing the door as far back from the outer elevation as possible and by providing a steeper fall to the balcony surface.

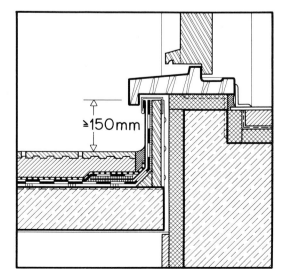

≥150mm

Balconies
Thresholds

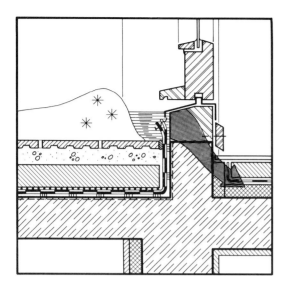

In some cases, although the sealing layer is raised at the threshold, rainwater penetrates behind its edge. The sealing end may be insufficiently covered by the edge trim or the overhang section, or the edge of the raised sealing layer is bonded only to the window section or fixed here and there with nails to the doorstep, so that it comes loose and slips down.
Sealing edges which are sealed with the aid of cement or mastic leak.

Points for consideration

— In wet weather water runs down the vertical external surfaces of buildings, e.g. doors and windows. This is driven by wind into cracks and hollows and can penetrate a sealing layer edge which is insufficiently covered.

— No lasting watertight construction is achieved if the raised sealing layer remains unprotected on the upper surface of the threshold. Sealing by means of cement or mastic has limited durability and requires constant maintenance.

— In balconies with lateral supports, e.g. cantilevers, strong movements of the balcony slab, caused by temperature stress and mainly parallel to the junction, can be expected at the connection between the edge of the slab and the threshold.

Recommendations for the avoidance of defects

● The edge of the sealing layer at the threshold must be protected from water flowing down the surfaces of doors and windows. This is achieved by placing it under the threshold section (pressed metal trim) or window ledge.

● If the balcony is constructed as a slab with lateral support, the sealing layer should be raised on the balcony slab itself and should be fixed independently of the façade. The waterproofing of the joint can be achieved by means of cover flashings.

● If, in balcony slabs with lateral supports, the angled sealing layer must be secured to the façade, the junction should be made with flexible synthetic foils.

● The angled sealing layer should be protected by the raised surface slabs or by sheet metal cladding suspended in front of it.

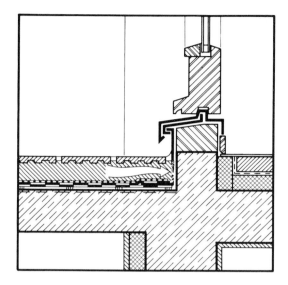

Where metal sheets or sections are used to form the junction of the sealing layer to the vertical threshold, leakages often result. In some cases, the junction of pressed metal flashings leaks because the sheet has not been carefully bonded between the layers of the sealing membrane, or has been bonded only on one side. Again, in the mortar bed of the top surface, metal sections, especially if of lead or zinc, that have been installed without a protective layer are chemically damaged by direct contact with the mortar particles, and become porous.

Points for consideration

— The use of sheet metal angles requires the permanent water-proof bonding of two very different structural materials with the aid of an adhesive; this demands special care.

— Metals, especially milled lead and zinc, are chemically vulnerable to fresh lime or cement mortar, in that the calcium hydroxide $Ca(OH)_2$ dissolves the metal. Direct contact must therefore be avoided and protective measures taken.

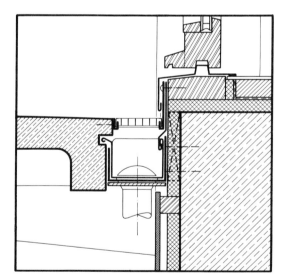

Recommendations for the avoidance of defects

● The angle and junction of the sealing layer to the threshold should be formed from the roof membrane itself or by the use of connection foils of similar material properties.

● The formation of the door threshold junction with pressed metal sections should be avoided if the threshold length exceeds 3 m.

● If metal sections are used, they should be given a protective coating (e.g. bitumen) to safeguard them from chemical reaction with fresh lime and cement mortar. To allow for linear movement, sliding seams should be provided at adequate centres (with titanium zinc, $\geqslant$ every 3 m). After the removal of dampness and dirt (oil, etc.), the sheet should be bonded in for at least 120 mm between the individual layers of the multi-layer membrane.

● The angled sealing layer must be protected by raising of the surface slabs or by freely movable sheet-metal claddings suspended in front of it.

Balconies
Thresholds

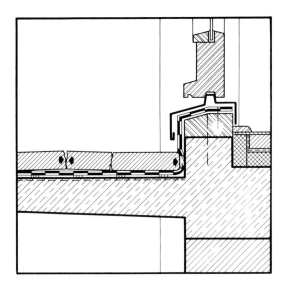

Where angles are formed from the sealing layer itself or by means of metal sections, leakages frequently occur due to mechanical stresses. At the change of plane the sealing layer may be deflected in a right angle and carried to the junction without a correspondingly shaped sealing base, so that it is unsupported and is pierced. In other examples the finishing surface is taken up to the angle (i.e. without a separating edge joint) and the sealing layer is damaged by surface movement.

Points for consideration

— During the construction period the unprotected sealing layer is subjected to strong mechanical stresses from pedestrian traffic, storage of building materials, etc.

— Multi-layer sealing layers and foils cannot be formed into a right angle. At right-angled junctions the bituminous adhesive layers crack and are unsupported. The sealing layer is particularly vulnerable to mechanical damage in places where it is unsupported.

— The surface layer is exposed to considerable changes in temperature and is therefore subjected to strong thermal linear expansion. The surface slabs are laid loosely on the sealing layer; this linear expansion takes effect at the edge of the continuous roofing surface, e.g. at the connections to vertical surfaces where the sealing layer is angled (see D 1.1.4 and 2.3.3 − Expansion joints in the balcony surface).

Recommendations for the avoidance of defects

● The upstand angle of the sealing layer to the surfaces of vertical structural members should be formed from the roof membrane itself or by the use of connection foils of similar material properties.

● At the edge the sealing layer should be raised without sharp deflections by using an edge fillet preformed in the sealing base; this guarantees a complete sealing support.

● The balcony surface must be separated from the angle by an edge joint taken as far as the separating layer.

General texts and principles

Balkowski, D.: Bautechnische Kriterien der Balkonplatte. In: Das Dachdeckerhandwerk (DDH), Heft 10/75, S. 696.
Moritz, Karl: Flachdachhandbuch – Flache und flachgeneigte Dächer, 4. Auflage, Bauverlag Wiesbaden und Berlin 1975.
DIN 4122 – Abdichtung von Bauwerken gegen nicht drückendes Oberflächenwasser und Sickerwasser mit bituminösen Stoffen, Metallbänder und Kunststoffolien, Richtlinien, Juli 1968.

Connection to vertical abutments

Forschungsgemeinschaft Bauen und Wohnen: Balkone im Wohnungsbau – Erfahrungen und Beobachtungen über Anordnung und Ausführung. Bericht der FBW, Heft 35/54, Stuttgart 1955.
Grün, Wolfgang: Balkone, Loggien, Terrassen. In: Deutsche Bauzeitschrift (DBZ), Heft 5/71, S. 937–950.
Lufsky, Karl: Bauwerksabdichtungen – Bitumen und Kunststoffe in der Abdichtungstechnik, 2. Auflage, B. G. Teubner, Stuttgart 1970.
Probst, Raimund: Außengänge, Balkone, Dachterrassen. In: Baupraxis, Heft 3/70, S. 39–43.
Zimmermann, Günter: Dachterrassen mit aufgesetzten Belägen. Durchfeuchtungen von Decken und Wänden. In: Deutsches Architektenblatt (DAB), Heft 12/74, S. 927.
Zimmermann, Günter: Plattenbeläge auf Balkonen. In: Architekt + Ingenieur, Heft 5/71, S. B1–B4.

Edge of structural member

Österreichisches Institut für Bauforschung: Abdichtungen und Abläufe bei Flachdächern, Dachterrassen, Balkonen, Loggien, Naßräumen, Forschungsbericht 39, 2. Auflage, Wien 1973.
Probst, Raimund: Außengänge, Balkone, Dachterrassen. In: Baupraxis, Heft 3/70, S. 39–43.
Schuster, Franz: Balkone – Balkone, Laubengänge und Terrassen aus aller Welt, Julius Hoffmann Verlag, Stuttgart 1962.

Expansion joints

Caemmerer, Winfried: Wärmeschutz aber richtig. Deutsches Bauzentrum, Köln 1958.
Eichler, Friedrich: Bauphysikalische Entwurfslehre, Band 2, 4. Auflage, Verlagsgesellschaft Rudolf Müller, Köln 1973.
Forschungsgemeinschaft Bauen und Wohnen: Balkone im Wohnungsbau – Erfahrungen und Beobachtungen über Anordnung und Ausführung, Bericht der FBW, Heft 35/54, Stuttgart 1955.
Grün, Wolfgang: Balkone, Loggien, Terrassen. In: Deutsche Bauzeitschrift (DBZ), Heft 5/71, S. 937–950.
Zimmermann, Günter: Plattenbeläge auf Balkonen. In: Architekt + Ingenieur, Heft 5/71, S. B1–B4.
AGI-Arbeitsblätter; A 10 – Industrieböden, Hartbetonbeläge; A 11 – Industrieböden, Zementestrich als Nutzboden.

Drainage

Österreichisches Institut für Bauforschung: Abdichtungen und Abläufe bei Flachdächern, Dachterrassen, Balkonen, Loggien, Naßräumen. Forschungsbericht 39, 2. Auflage, Wien 1973.
Probst, R.: Außengänge, Balkone, Dachterrassen. In: Baupraxis, Heft 3/70, S. 39–43.
Schuster, F.: Balkone – Balkone, Laubengänge und Terrassen aus aller Welt, Julius Hoffmann Verlag, Stuttgart 1962.
DIN 1986 – Grundstücksentwässerungsanlagen, technische Bestimmungen für den Bau, Juni 1962.

Thresholds

Grün, Wolfgang: Balkone, Loggien, Terrassen. In: Deutsche Bauzeitschrift (DBZ), Heft 5/71, S. 937–950.
Lufsky, Karl: Bauwerksabdichtungen – Bitumen und Kunststoffe in der Abdichtungstechnik, 2. Auflage, B. G. Teubner, Stuttgart 1970.
Probst, Raimund: Außengänge, Balkone, Dachterrassen. In: Baupraxis, Heft 3/70, S. 39–43.
Zentralverband des Dachdeckerhandwerks: Richtlinien für die Ausführung von Flachdächern, Ausgabe Januar 1973, Helmut Gros Fachverlag, Berlin 1973.

Key to symbols used in the illustrations

Note: The illustrations represent principles and are not always to scale

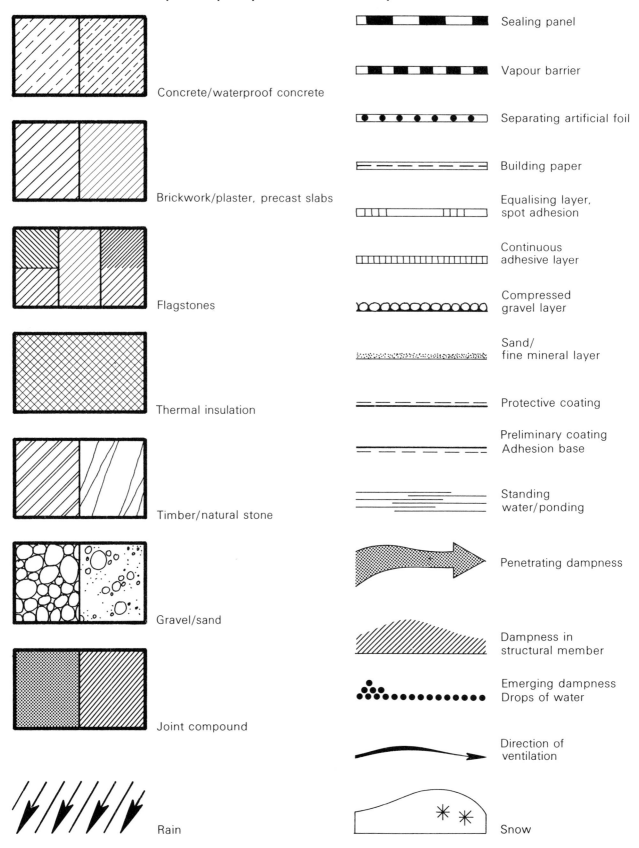

Concrete/waterproof concrete

Brickwork/plaster, precast slabs

Flagstones

Thermal insulation

Timber/natural stone

Gravel/sand

Joint compound

Rain

Sealing panel

Vapour barrier

Separating artificial foil

Building paper

Equalising layer, spot adhesion

Continuous adhesive layer

Compressed gravel layer

Sand/fine mineral layer

Protective coating

Preliminary coating Adhesion base

Standing water/ponding

Penetrating dampness

Dampness in structural member

Emerging dampness Drops of water

Direction of ventilation

Snow

English language bibliography

Addleson, L. (1977) Guide to building failures. Flat roofs. *Architects' Journal* **165**(7), 321–7; **165**(9), 415–21; **165**(13), 615–7.

Architects' Journal (1972) Flat roofs: appalling experience with Crown buildings. **154**(36), 521–2.

Architects' Journal (1976) Repair and Maintenance. Technical Study 3. Flat roofs. **163**(25), 1239–44.

Bickerdike, E. (1973) Car park roof. *Building* **225**(6812), 60.

Bickerdike, Allen, Bramble with O'Brien, T. (1974) Prestressed concrete roof. *Building* **227**(6850(38)), 123; Edge of asphalt roof. *Building* **227**(6856(44)), 79.

Bickerdike, Allen, Rich and Partners with O'Brien, T. (1973) Roof with heated ceiling. *Building* **224**(6733(12)), 116; Asphalt roof. *Building* **225**(6809(48)), 87.

Borer, K. (1974) Flat roofs. *RIBA Journal* **81**(6), 37–43.

British Standards Institution, London

British Standards:

BS 747: Part 2: 1970 *Roofing felts.*

BS 1162, 1418, 1410: 1973 *Mastic asphalt for building (natural rock asphalt aggregate).*

BS 2972: 1975 *Methods of test for inorganic thermal insulating materials.*

BS 3533: 1962 *Glossary of terms relating to thermal insulation.*

BS 5250: 1975 *Code of basic data for the design of buildings: the control of condensation in dwellings.*

British Standard Codes of Practice:

CP3: Chapter II: 1970 *Thermal insulation in relation to control of the environment.*

CP3: Chapter VIII: 1949 *Heating and thermal insulation.*

CP 110: Part 1: 1972 *Structural use of concrete,* Appendix B.

CP 143: Part 16: 1974 *Semi-rigid asbestos bitumen sheet.*

CP 144: *Roof coverings*

Part 3: 1970 *Built-up bitumen felt.*

Part 4: 1970 *Mastic asphalt.*

CP 308: 1974 *Drainage of roofs and paved areas.*

Broderick, A. H. Y. (1960) Condensation under impervious roofing. *The Builder* **191**(6131), 943–6.

Building (1976) Leaks in single-coat asphalt. **230**(6930(16)), 81; Deflection in concrete; **230**(6935(21)), 111.

Building Research Establishment (1970) *Built-up Felt Roofs.* Digest No. 8. Garston.

Building Research Establishment (1975) *Condensation in Roofs.* Digest No. 180. Garston.

Building Research Station (1961) *Principles of Modern Building,* vol. 2 *Floors and Roofs.* London, HMSO.

Burns, J. U. (1975) Trends in flat roofing. *Construction* (15), 30–33.

Cayley, D. (1974) Reducing roof maintenance by design. *The Architect* **4**(9), 60–62.

Day, A. G. (1975) Technical Study: The inverted roof. *Architects' Journal* **161**(2), 1047–52.

Edwards, R. M. (1962) Thermal insulation: roofs and problems associated with moisture. *Architects' Journal* **136**(18), 1029–30.

Eldridge, H. J. (1976) *Common Defects in Buildings.* London, HMSO.

Freeman, I. (1975) Building failure patterns and their implications. *Architects' Journal* **161**(6), 303–08.

Handeford, G. O. (1964) *Problems in Flat Roofs: A Review of Research.* Technical Paper 182, Research Paper No. NRC 8024, Division of Building Research, National Research Council.

Handisyde, C. (1967) Building enclosure: roofs. Technical study. Flat roof failures: causes, and how to avoid them. *Architects' Journal* **146**(26), 1659–70.

Jain, S. P. and Rao, K. R. (1974) Movable roof insulation in hot climates. *Building Research and Practice* **2**(4), 229–34.

Johnson, D. K. (1972) Recommendations for flat roof coverings. *Building* **223**(6753), 120–24.

Latham, J. P., *et al.* (1972) *Asphalt and Built-up Felt Roofings: Durability.* Digest No. 144, Building Research Station, Garston.

May, K. O. and Johnson, D. K. (1973) *Flat Roofs – Design, Maintenance and the Use of Plastics.* Interbuild Conference, Olympia, London, 15 November.

McInnes, H. W. (1973) Designing the trouble-free roof. *Building* **224**(6769), 116.

Murray, J. (1971) Flat roof failures. Parts 1 and 2. *Architects' Journal* **153**(26), 1489–93; **154**(1), 37–41.

Plonski, W. (1971) Results of heat–moisture tests of flat roofs of various types. *Building Science* **6**(1), 1–6.

Scott, G. (1976) *Building Disasters and Failures: A Practical Report.* Hornby, Lancs, Construction Press.

Sengler, D. (1973) *Flat Roofs.* Translation 1977, Building Research Establishment Library, Garston.

Vos, B. H. (1971) Condensation in flat roofs under non-steady-state conditions. *Building Science* **6**(1), 7–15.

Zimmerman, R. (1975) Development of the upside-down roof. *Insulation* **19**(5), 11–13.

Index